Introduction

This manual consists of a variety of experiments, all of which illustrate biological concepts using micro-organisms as the experimental material. The title of the manual includes the word 'micro-organisms', therefore we should begin by defining what we mean by this term.

Microbiology is the branch of biological science that deals with a heterogeneous, i.e. mixed group of organisms which we call **micro-organisms.** Micro-organisms are normally grouped into a Kingdom—the **Protista,** which includes bacteria, fungi, slime moulds, algae and protozoa. This grouping is largely one of convenience but is justified since all the organisms have a number of features in common, namely:

1. They are very small, and to determine structural details, most of the Protista need to be studied under a microscope. The only exceptions are some marine algae, which grow to a large size. These organisms, however, show obvious relationships to the other members of the Kingdom.
2. They all show a simple structure.
3. Techniques for handling and culturing these organisms in the laboratory are essentially the same for all micro-organisms.

The study of micro-organisms in detail may seem from these features to be somewhat difficult and unrewarding, but this is far from true as there are many reasons why we should study members of this group.

Firstly, micro-organisms are of considerable value in experimental biology. They show all the basic characteristics that we normally associate with living organisms but they have the advantages that: (*a*) They are usually easy to handle under experimental conditions in the laboratory, and (*b*) whole populations can be studied in a very short time. Secondly, much of our knowledge of cell metabolism and

1

genetics has come from research which used micro-organisms.

A third and most significant reason is that micro-organisms are of considerable economic importance to man in both advantageous and disadvantageous roles. Some of the ways in which micro-organisms affect us are listed below.

In industry	In agriculture	In public health	Causing biodeterioration
Production of:	Role in soil	Human disease	Of paint
organic acids	ecology	Food poisoning	wood
antibiotics	Causing plant	Sewage disposal	cotton
vitamins	disease	Water	foods
enzymes	Causing animal	purification	plastics
foods	disease	Quarantine	
alcoholic drinks			
and in tanning			

It can be seen that there are very few, if any, aspects of man's life, which are not influenced either directly or indirectly by micro-organisms.

However, in order to recognise individuals they must be given a name.

NAMING AND CLASSIFYING MICRO-ORGANISMS

Naming micro-organisms

Giving a name to a living organism is essentially a method of identifying it so that the relevant information can be conveyed about the organism.

In all languages, trivial (common) names are given to familiar living organisms. Not all organisms, however, are named, particularly in the case of micro-organisms, and the trivial name will be different in each language and may even vary from one district to another within one country. For these reasons trivial names are very rarely scientifically accurate. It is essential, however, if information is to be conveyed internationally, that there should be an internationally recognisable **nomenclature,** i.e. a system of naming organisms.

The present accepted nomenclature is called the **binomial system.** It was introduced by Carl von Linné (Linnaeus), a Swedish botanist, who was the first scientist to attempt any comprehensive grouping of living organisms. In this system each distinct kind of organism is called a **species** and is given a specific name. The complete name of the organism, however, consists of two words, the first being the

EXPERIMENTAL BIOLOGY WITH MICRO-ORGANISMS

STUDENTS' MANUAL

J. W. GARBUTT, M.SC., L.I.BIOL.
Lecturer in Microbiology,
Grimsby College of Technology

A. J. BARTLETT, B.SC.
Head of Science,
Little Heath Comprehensive School, Reading

LONDON
BUTTERWORTHS

THE BUTTERWORTH GROUP

ENGLAND

Butterworth & Co (Publishers) Ltd
London: 88 Kingsway, WC2B 6AB

AUSTRALIA

Butterworth & Co (Australia) Ltd
Sydney: 586 Pacific Highway Chatswood, NSW 2067
Melbourne: 343 Little Collins Street, 3000
Brisbane: 240 Queen Street, 4000

CANADA

Butterworth & Co (Canada) Ltd
Toronto: 14 Curity Avenue, 374

NEW ZEALAND

Butterworth & Co (New Zealand) Ltd
Wellington: 26-28 Waring Taylor Street, 1

SOUTH AFRICA

Butterworth & Co (South Africa) (Pty) Ltd
Durban: 152-154 Gale Street

First published 1972

© Butterworth & Co (Publishers) Ltd, 1972

IBSN 0 408 70227 3 Standard
      0 408 70228 1 Limp

Suggested U.D.C. number 576·8·001·5
Suggested additional number 578·08·576·8

Filmset by V. Siviter Smith & Co Ltd, Birmingham
Printed in England by J. W. Arrowsmith Ltd, Bristol

Contents

name of the **genus** (generic name), or group to which the organism belongs, and the second word being the specific name. The generic name is always treated as a Latin noun, even though it may be a latinised Greek word, or a word made up from Latin or Greek stems, or even a word originating in another language. The genus always has a capital letter. The specific name, however, never begins with a capital letter, and is often a Latin adjective, but may be a noun which modifies or explains the generic name.

There exists an *International Code of Nomenclature* so that when new scientific names are created they are internationally accepted.

Mostly the names of micro-organisms describe the morphology of the organism, or else relate to the effect the organism has, or to the scientist who discovered it. Some examples are quoted below:

Generic name		*Specific name*	
Mycobacterium	(fungus rodlet)	*leprae*	(of leprosy)
Bacillus	(a rodlet)	*cereus*	(wax-coloured)
Clostridium	(a small spindle)	*welchii*	(after Welch, the scientist who first isolated it)
Escherichia	(after Prof. Theodor Escherich)	*coli*	(of the large intestine)

CLASSIFYING MICRO-ORGANISMS

There are many thousands of different types of living organisms in existence today. Even conservative estimates give the numbers of different kinds of bacteria as 1500 species, of fungi as 80 000, and of protozoa as 45 000, and these figures only refer to the numbers that have been described. These lists are constantly being added to so there may be many thousands more that scientists have not yet seen. Thus the need for organisation into manageable units is obvious, for no human being is capable of even glancing at such a vast array of living things, let alone studying them in any detail.

How can we begin to understand micro-organisms as a whole? The answer seems to lie in the way in which the human brain normally operates, i.e. it tends to sort information into manageable categories rather than retaining isolated facts. One can draw an analogy with the situation of the goods in a supermarket. If they are arranged in an organised way, e.g. meats together, tinned fruit together, etc., then shopping presents no problem, but imagine the chaos if the goods were arranged randomly!.

To satisfy this need, scientists have evolved systems for grouping living organisms, so that the information relating to each group can be handled with relative ease. For example:

There are 118 different kinds of bacteria all of which have the following characteristics:

(*a*) They are Gram-positive (see Experiment 3.1, page 57).
(*b*) They are rod-shaped.
(*c*) They produce endospores.

These bacteria are placed in a group called the Bacillaceae. If we encounter an organism which we know belongs to this group, even without seeing it or studying it we know it exhibits the features listed above.

The placing of living things into groups is called **classification**. The branch of science that deals with classification, and the description and naming of organisms, we call **Taxonomy**. The vast range of organisms necessitates a fairly complex filing system and so taxonomists have devised a number of taxa (singular—taxon) or subdivisions. The main taxonomic categories are:

species	(plural species)	Comprising organisms of one type
Genus	(plural Genera)	Comprising related species
Family	(plural Families)	Of related genera. (*Note*. Family names usually end in -aceae)
Order	(plural Orders)	Of related families. (Order names usually end in -ales)
Class	(plural Classes)	Of related orders. (Class name usually end in -cetes)
Phylum	(plural Phyla)	Of related classes. (Names usually end in -phyta)
Kingdom	(plural Kingdoms)	Of related phyla

Other sub-divisions, for example sub-order, sub-family, sub-class, etc., may be needed for some groups of organisms due to more complex interrelationships. In order to see how the system operates consider the following examples:

The tubercle bacillus is an organism causing the serious disease of tuberculosis in man. This organism is considered to be an individual distinct from other bacteria and is therefore a separate species. The scientific name for this organism is *Mycobacterium tuberculosis*. A number of bacteria are considered to be closely related to *M. tuberculosis* and so they are given the same generic name—*Mycobacterium*. These constitute the genus *Mycobacterium* of which there are 14 members, for example *M. leprae* (causing leprosy), and *M. phlei* (a harmless soil bacterium).

The genus *Mycobacterium* is related to another genus, *Myococcus*. The organisms in the two groups only differ in their response to a particular staining reaction, and in the fact that *Myococci* are pigmented whereas *Mycobacteria* are not. Both genera are included in the family **Mycobacteriaceae**.

Members of the *Mycobacteriaceae*, together with the families *Actinomycetaceae* and *Streptomycetaceae* constitute the **Order Actinomycetales**. All organisms in this order show some degree of mycelial branching, resembling fungi, and all are Gram-positive (see Experiment 3.1).

All order of bacteria are grouped together under the **Class Schizomycetes**. (See *Table I.2* on page 7.) The bacteria closely resemble the blue-green algae (**Class Schizophyceae**) with regard to basic cell structure. Thus they are included in the **Phylum Schizomycophyta** (*Procaryota*) a group often called the Lower Protista (see *Tables I.1* and *I.2*).

The classification of the tubercle bacillus is as follows:

Kingdom	:	Protista
Phylum	:	Schizomycophyta
Class	:	Schizomycetes
Order	:	Actinomycetales
Family	:	Mycobacteriaceae
Genus	:	*Mycobacterium*
Species	:	*tuberculosis*

Having now considered the binomial system in principle and an example of its usage, let us turn to consider the basis for the classification of a particular group of organisms.

It is unfortunate that there is no one universally accepted system of classification for any group of organisms. This is due to the fact that taxonomy does not aim only to produce a filing system, but taxonomists also try to show by their classificatory systems the interrelationships of organisms in the living world. In addition, the systems attempt to reflect the evolutionary history of an organism in relation to the group to which it belongs. A system which has these aims is called a **natural** system in comparison to the purely filing system type or **artificial** system.

Most classificatory systems tend to be a mixture of natural and artificial systems, simply because we often know very little about the past evolutionary history of a particular species. We therefore have to rely on present-day characteristics. Which of these should be used? Firstly, as much detailed knowledge as possible should be gained about ALL the organisms involved, but even then the various experts may disagree about which characteristics should be used for classification or which are the most important. Hence the existence of the various taxonomic systems.

Table I.1 GENERAL CLASSIFICATION OF MICRO-ORGANISMS

Viruses	Protista	
	Lower Protists (*Procaryota*)	*Higher Protists* (*Eucaryota*)
Organisms consisting of a nucleic acid core surrounded by a protein coat.	Nuclear material not bounded by a nuclear membrane	Nuclear material separated from cytoplasm by a nuclear membrane
Only one type of nucleic acid is present, i.e. either DNA or RNA, but not both.	Cytoplasm structurally simple	Cytoplasm structurally complex
Only reproduce inside living host cells.	No mitochondria	Mitochondria present
	Pigments when present not in plastids.	Plastids present in some cells
	No cytoplasmic streaming	May show cytoplasmic streaming
	Flagellae simple; resemble a single filament of a eucaryotic flagellum	Cilia or flagellae, where present, show a 9:2 fibrillar structure
	Bacteria	Protozoa
	Blue-green algae	Algae (except Blue-greens)
		Fungi
		Myxomycetes

Note: All of the above characteristics are also common to higher plants and animals.

Table I.2 THE SUBDIVISIONS OF THE LOWER PROTISTS

Procaryota (Lower Protists)

	Schizomycetes (Bacteria)	*Schizophyceae* (Blue-green algae)
		Photosynthetic organisms Cells often in chains or filaments Contain the pigment phycocyanin Gliding motility
Eubacteriales.	'True Bacteria' Simple spherical or rod-shaped rigid cells Not photosynthetic. Stain easily with basic aniline dyes Motility (where present) via peritrichous flagella	
Pseudomonadales.	Rod-shaped or curved or spiral bacteria If motile, then via poor flagella. All Gram −ve Some photosynthetic types	
Actinomycetales.	Filamentous bacteria. Gram +ve. Includes *Mycobacterium tuberculosis*; also *Streptomyces* sp. producing streptomycin	
Spirochaetales.	Helical organisms with flexible cells Move by contraction not flagella	
Myxobacteriales.	Slime bacteria Undergo aggregation to give a complex fruiting structure	
Chlamydobacteriales.	Filamentous iron bacteria	
Beggiatoales.	Filamentous sulphur bacteria	
Rickettsiales.	Small bacteria-like organisms causing typhus, and Q-fever in man, using Arthropod vectors Like viruses, cannot reproduce outside living host cells	

Table I.3 THE SUBDIVISION OF THE HIGHER PROTISTS

Group	Subdivision	Characteristics
Myxomycophyta. Slime moulds. These organisms have a vegetative phase consisting of a multinucleate mass of protoplasm—the plasmodium. Other phase is one of single cells—myxamoebae, or flagellate swarmers		
Eumycophyta. Fungi. Unicellular or filamentous organisms, non-photosynthetic but with definite cell walls	Phycomycetes.	Water moulds, downy mildews and mucors. Mycelium normally non-septate. Asexual spores produced in sporangia. Resting spore/sporangia produced sexually.
	Ascomycetes.	Cup fungi, flask fungi, powdery mildews, yeasts, blue and green moulds. Hyphae septate, sexual reproduction produces asci plus 8 or occasionally 4 ascospores.
	Basidiomycetes.	Mushrooms, toadstools, bracket fungi, puffballs, stinkhorns, smuts and rusts. Hyphae septate, sexual reproduction gives basidia with four externally arranged basidiospores
	Fungi Imperfecti.	Fungi without a sexual stage. Usually producing asexual conidia
Protozoa. Acellular (or unicellular) animals. Some groups contain organisms that could easily be classified as Algae.	Rhizopoda.	Amoebae, sun animalcules, foraminifera and radiolarians. Main phase of life cycle amoeboid
	Flagellata.	Flagellates. Main phase with one or more flagellae
	Ciliophora.	Ciliates. Plus cilia as locomotory organs
	Sporozoa.	Parasitic protozoa. Intracellular, no locomotory structures, e.g. malaria parasite
Algae. Mainly aquatic organisms, or live in damp habitats. Photosynthetic, containing the pigment chlorophyll a, found in higher plants. No conducting system. A heterogeneous group therefore split into six phyla.	Chlorophyta.	Green algae. Marine and freshwater algae with chlorophyll a and b, as in higher plants. Unicellular, filamentous or multicellular. Store starch or oil.
	Chrysophyta.	Diatoms and related forms. Freshwater/marine. Unicellular or multicellular. Contain pigments chlorophyll a and sometimes c, also the yellow pigment xanthophyll. Cell wall often contains silica.
	Pyrrophyta.	Dinoflagellates. Unicellular, flagellate. (Sometimes included in the Protozoa.) Contain chlorophyll a and c and special carotenoids.
	Phaeophyta.	Brown algae. Marine, multicellular, containing chlorophyll a and c, plus special pigments. Some highly specialised forms
	Euglenophyta.	Euglenoids. Freshwater, unicellular, flagellates. Contain chlorophyll a and b, but store paramylum starch, and fat. (Often included in the Protozoa)
	Rhodophyta.	Red algae. Usually multicellular, containing chlorophyll a, and special pigments phycoerythrin and phycocyanin

A new approach has recently been employed in an attempt to overcome the objective nature of the classical method. As many characters as possible are determined for each organism in a particular group. Each character is given equal weight and the organisms are compared using computer analysis. By this method the organisms most closely related are those with the largest number of features in common.

The diagrams, *Tables I.1*, *I.2*, and *I.3*, show an outline classification of the major groups of micro-organisms. Whenever a new micro-organism is studied try to see where it fits into the scheme and how it is related to other micro-organisms.

1
Handling, culturing and observing micro-organisms

INTRODUCTION

The basic techniques for handling, culturing, and observing micro-organisms, were developed many years ago, mainly by scientists working with bacteria. Although technical advances have since been made, the mastery of these basic techniques is still essential to successful experimentation with micro-organisms.

Most micro-organisms are very small, and some are disease-causing organisms—**pathogens**. These latter must therefore be handled safely as well as efficiently.

The experiments in this chapter, if followed carefully, will enable you to handle micro-organisms in the laboratory with safety and efficiency. As you use the techniques, think about how they prevent the escape of micro-organisms from dishes, tubes and flasks. How do they prevent unwanted organisms from entering and interfering with your experiments?

These experiments are not only to teach you technique, however, as they involve aspects of the distribution, size and shape, and resilience of micro-organisms. Do not therefore lose sight of these, the major aspects of the experiments.

EXPERIMENT 1.1

MICRO-ORGANISMS—THEIR VARIETY AND THEIR HABITATS

THEORY

In this first experiment you will investigate a number of different situations, which we can call environments, to see whether micro-organisms are present or not. Their presence or absence will depend very largely on whether they are able to obtain food materials for growth and reproduction from these environments. Thus, if we are to grow micro-organisms in the laboratory, we must provide the food materials that these living organisms need.

Micro-organisms are grown in the laboratory in solutions which are known as **media**. When the media are liquid and made from meat extracts or plant extracts they are often called **broths**. Media can either be liquid or solid, the only difference being that the solid media contain an additional compound called agar. This is an inert carbohydrate extracted from certain seaweeds. It has the interesting property of producing a hydrophilic colloid, the **gel** form of which does not become a **sol** until heated to 98°C, and yet remains as a sol until cooled to 35–40°C (see below):

Room temperature increasing to 98°C
GEL (solid ————————————→SOL (liquid)
       remaining as gel until

       decreasing temp.
. . . GEL ←———————————— SOL
at 35–40°C   remains as sol until

PROCEDURE

Part 1. Pouring the plates for the experiment

1. Take a sterile Petri dish and put it on your bench.
2. Light your Bunsen and adjust the air-hole to give you an all-blue flame.
3. Take a bacteriological tube or MacCartney bottle (see *Figure 1.1a*) containing 15 cm³ of sterile molten (i.e. sol form) agar, from the 45°C water-bath.
4. Wipe the outside of the tube containing the agar with a paper tissue and remove the cap from the tube.
5. Flame the mouth of the tube, i.e. very quickly pass the mouth of the tube through the Bunsen flame, rotating the tube whilst it is in the flame.

DO NOT HOLD THE TUBE IN THE FLAME FOR MORE THAN A FEW SECONDS.

6. Using your left hand, tilt the lid of the sterile Petri dish, just enough to introduce the mouth of the bacteriological tube (see *Flow Diagram 1.1*).
7. Pour the agar into the dish quickly, making sure that the tube does not touch the sides of the dish or lid.
8. Withdraw the tube and close the lid.

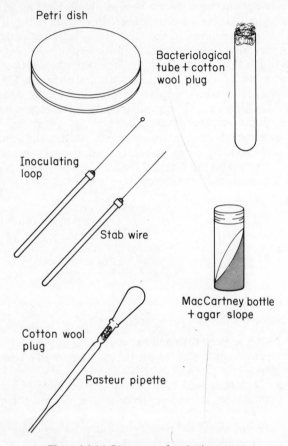

Petri dish

Bacteriological
tube + cotton
wool plug

Inoculating
loop

Stab wire

MacCartney bottle
+ agar slope

Cotton wool
plug

Pasteur pipette

Figure 1.1 (a) *Diagram to show basic apparatus*

9. With the Petri dish FLAT on the bench, gently rotate the dish
   so that the agar forms an even layer at the bottom of the Petri
   dish.
10. Allow the agar to cool and set.
11. Invert the dish to prevent any condensation falling on the agar
    surface. The completed Petri dish plus agar is called a plate.
12. Label the bottom of the dish (using a wax pencil or felt-tipped
    pen) with the type of agar, for example for nutrient agar use the
    letters NA, your group number, and the date.
13. Each group should pour eight nutrient agar (NA) plates and
    three malt extract agar (MA) plates.

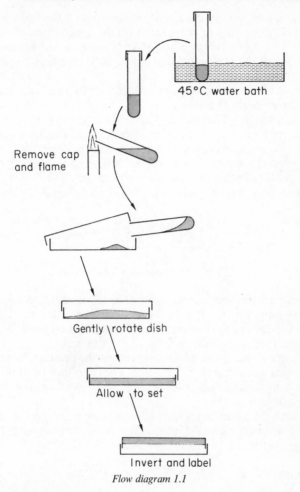

45°C water bath

Remove cap
and flame

Gently rotate dish

Allow to set

Invert and label

Flow diagram 1.1

Part 2. Sampling various environments to see if micro-organisms are present or absent

1. Take one NA plate and one MA plate prepared as above and expose them to the air in the laboratory for 30 min. Leave the dishes and lids as shown in *Figure 1.1 (b)* below. Repeat this and expose an NA plate and an MA plate to the outside air for 30 min.
2. Carry out the following with NA plates only:
   (*a*) Touch the surface of an agar plate with a finger.
   (*b*) Place a hair on the agar surface.
   (*c*) Place some finger-nail scrapings on the plate.

    (*d*) Sprinkle a small amount of fresh soil on the surface.

    (*e*) Place a drop of pond water on the agar surface.

    (*f*) Leave one NA and one MA plate untouched as controls.

3. Label all the plates to show where they have been exposed or what they have on them.

4. Leave all the plates at room temperature for one week and then examine them as follows.

    Look at the plates exposed to the air. Micro-organisms usually occur as more or less discrete, circular areas known as **colonies**. The assumption is made that each colony has been produced by a single bacterium, yeast cell or fungal spore, falling on the plate. By asexual reproduction, this single cell will give rise to many thousands of individuals which make up the colony.

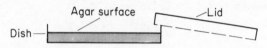

Figure 1.1 (b) *To show plate exposed to air*

5. Describe the size, shape and colour of the colonies. Sketches may help you to do this more accurately. Also describe the consistency of the colonies, i.e. whether they are dry or slimy in appearance; smooth, or fluffy like cotton wool.

6. Examine your other plates and record the presence or absence of micro-organisms. Discrete colonies may not be apparent if large numbers of micro-organisms are present causing colonies to overlap.

7. Count the colonies on the plates that were exposed to ⸱he air. Calculate the rate of precipitation of micro-organisms on the plate per metre of surface per hour.

QUESTIONS

1. From the results of your experiments do you think that micro-organisms are widely distributed or confined to a number of very specific habitats?

2. Do you think that the organisms growing on your plates are all of the same kind? If not, give reasons for your answer.

3. Why do you think agar is used as a geling agent in media?

4. Why is it important to prevent any condensation falling on the agar surface?

5. The presence of unwanted organisms in a culture is known as contamination and the organisms are called contaminants. How do the processes of (1) wiping the tube containing the sterile

agar, and (2) flaming the mouth of the tube, prevent contamination of the agar when pouring the plate?
6. What is the purpose of the control plates?

EXPERIMENT 1.2

ISOLATING A PURE CULTURE OF BACTERIA FROM A MIXED POPULATION

THEORY

In Experiment 1.1 colonies of micro-organisms were obtained by
exposing plates to the air, and the assumption was made that these
colonies arose from a single individual by asexual reproduction. This
may not be true as two different organisms may have fallen very
close to one another on the plate and the resulting colony, though
seeming to be a single colony, in fact would consist of a mixed
population.

Cultures of micro-organisms from natural sources often contain
mixed populations and in order to study an environment micro-
biologists usually start by separating the component organisms.
These are then studied separately.

In the following experiment you will use two different techniques
in an attempt to separate the two components of a mixed population.
This population contains a red bacterium—*Serratia marcescens*, and
Sarcina lutea, a yellow bacterium.

PROCEDURE

Part 1. Producing a series of dilutions

You will be provided with a nutrient broth culture which has been
growing overnight, containing the two organisms, *Serratia* sp. and
Sarcina sp.

1. Take the broth culture, an inoculating loop and a Bunsen, to
   your bench. Also take two bacteriological tubes each containing
   10 cm³ of sterile distilled water.
2. Light the Bunsen and adjust it to give a blue flame.
3. Hold the broth culture tube in your right hand and shake it to
   disperse the organisms evenly throughout the broth (see *Flow
   Diagram 1.2(1)*.
4. Transfer the tube to your left hand.
5. Sterilise the inoculating loop by holding it in your right hand
   between thumb and first finger like a pen, and place it in the
   Bunsen flame almost vertically so that the whole of the loop
   glows red hot (see *Figure 1.2* on page 17).

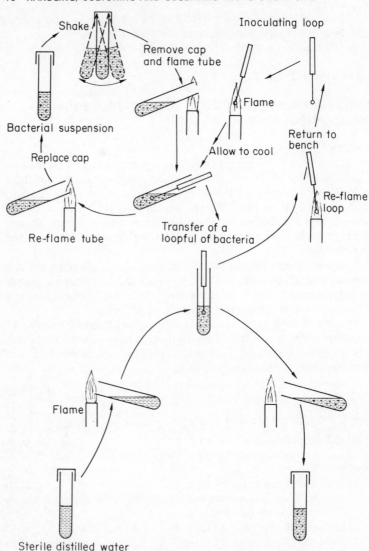

Flow diagram 1.2(1)

6. Retain the loop in your right hand. Hold the culture tube at an angle of 45 degrees and put the cap end of the tube into the palm of your right hand. Curl the little finger of your right hand round the cap of the tube and remove the cap in such a way as not to contaminate the edge or the inside of the cap (see *Figure 1.4*).

Figure 1.2 *Flaming an inoculating loop*

7. Flame the top of the tube.
8. Introduce the sterile loop into the tube and pick up a loopful of the broth culture. Try not to touch the side of the tube with the loop.
   Note. AFTER FLAMING THE LOOP MUST COOL FOR 15–20 s BEFORE YOU PUT IT INTO THE CULTURE.
9. Flame the neck of the tube and replace the cap. Put the tube back in its rack.
10. Take a bacteriological tube with cap containing 10 cm³ of sterile water. Using the technique described in Stages 4–9 open the tube and put the loopful of broth into this tube. Label this dilution A.
   DO NOT FORGET TO FLAME THE MOUTH OF THE TUBE.

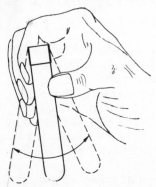

Figure 1.3 *Shaking an inoculated culture*

11. Now flame the loop and put it on the bench.
12. Shake the diluted culture as shown in *Figure 1.3*.
13. Repeat the whole procedure and obtain a further dilution; i.e. take a loopful of dilution A and put it into another tube containing 10 cm³ of sterile distilled water. Label this dilution B.
14. Repeat Stage 13 and by using dilution B obtain a further dilution—label this dilution C.

ALWAYS USE THESE TECHNIQUES WHEN HANDLING BROTH CULTURES OR LIQUID MEDIA IN TUBES (or MacCARTNEY BOTTLES).

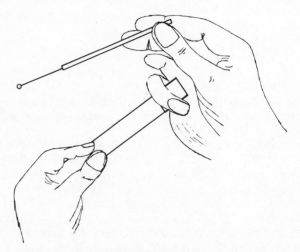

Figure 1.4 *The removal of a bacteriological tube cap*

Part 2. Separating the bacteria by streaking out

1. Using the technique learnt in Experiment 1.1, pour four nutrient agar plates.
2. Allow these to set and place them lid upwards on the bench.
3. Using the technique described in Part 1, remove a loopful of the original broth suspension.
4. Lift the lid of the Petri dish as though you were pouring agar into it, i.e. use your left hand and only raise the lid slightly
5. Introduce the loop and hold it against the agar surface on the opposite side of the plate (see *Flow Diagram 1.2(2)*). Hold the loop lightly and DO NOT press down, otherwise you will dig into the agar.
6. Lightly streak the loop across the plate as shown in *Figure 1.5* below. Lift the loop at the end of each streak line.

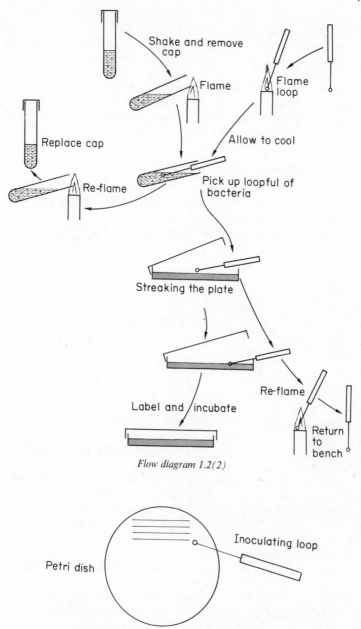

Shake and remove cap

Flame

Flame loop

Replace cap

Allow to cool

Re-flame

Pick up loopful of bacteria

Streaking the plate

Re-flame

Label and incubate

Return to bench

Flow diagram 1.2(2)

Inoculating loop

Petri dish

Figure 1.5 *To show first stage of plate streaking*

7. Close the dish and re-flame the loop. Allow the loop to cool.
8. Take the dish and loop and complete the streaking as shown in *Figure 1.6*. Rotate the dish, but maintain the lid in a semi-open state, as in *Flow Diagram 1.2(2)*, and preserve the aseptic technique.
9. Close the lid and re-flame the loop.
10. Repeat stages above for dilutions A, B, and C, thus preparing four streak plates, one for each dilution. Label the plates Experiment 1.2, and with your Group No. and the dilution letter.

Part 3. Separation by the pour plate method

1. Take a sterile Pasteur pipette. (Note the cotton wool plug at the end of the pipette.) Put a rubber teat on the pipette. Light your Bunsen and obtain a blue flame.
2. Take a tube of agar from the 45°C water-bath and stand it in your test-tube rack.

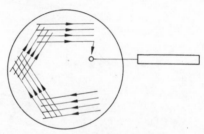

Figure 1.6 *Completed streak plate*

3. Hold the pipette using the thumb and first two fingers of your right hand. Thoroughly shake the suspension (dilution) C prepared above and using the little finger of your right hand, remove the cap.
4. Pass the pipette quickly through the Bunsen flame and flame the mouth of the tube containing the suspension.
5. Suck up a small amount of the suspension into the pipette. CAUTION—Do not compress the teat too much or liquid may be sucked into the cotton wool or even the teat.
6. Re-flame the tube and replace the cap.
7. Take the agar tube, remove the cap and flame the mouth of the tube.
8. Discharge five drops of suspension into the agar, re-flame the tube, replace the cap and put the tube back into the test-tube rack.
9. Return any residual liquid from the pipette back to its tube.

10. Discard the Pasteur pipette into a jar of disinfectant.
11. Mix the agar and bacteria (produced in Stage 8 above), by rotating the tube between the palms of your hands several times.
12. Pour the agar-bacteria mixture into a sterile Petri dish.
13. Repeat Stages 2–12 with suspension B, A and the original broth suspension.
14. Allow the plates to set. Invert them and label them Experiment 1.2, Part 3, and with your Group No. and the dilution factor.
15. Leave the plates at room temperature for a few days.
16. Examine the plates and describe their appearance.

Part 4. Isolation of a bacterium and production of a pure culture

For future work it is necessary to isolate these bacteria and grow them in a **pure culture**—a culture containing only one bacterial species or strain.

1. Carefully examine the plates from Parts 2 and 3 and select a yellow colony, isolated as far as possible from adjacent colonies.
2. Mark the position of the colony on the back of the plate using a wax pencil (or felt-tip pen).
3. Take a MacCartney bottle containing nutrient agar solidified as a slope (see *Flow Diagram 1.2(3)*).
4. Take a straight wire and flame-sterilise it by holding it in a Bunsen flame as you did the loop (see *Flow Diagram 1.2(3)*). Keep the wire in your right hand.

The introduction of micro-organisms into a culture medium is called **inoculation.**

5. Take the plate with the chosen colony and remove the lid. Hold the plate vertically in your left hand and stab the colony with the sterile wire. Do not try to remove the complete colony, simply stab the wire in and out again once. Replace the lid of the dish.
6. Take the MacCartney bottle in your left hand and remove the screw cap by holding the cap against the palm of your right hand and curling the little finger of your right hand round it. Then rotate the bottle using your left hand and so unscrew it. (See *Figure 1.4* for the position using a bacteriological tube—essentially similar to the technique for a MacCartney bottle.) Hold the bottle as near to the horizontal as possible.
7. Flame the neck of the bottle.
8. Inoculate the slope by drawing the wire lightly across the agar slope starting at the bottom of the bottle and working towards the neck.
   DO NOT DIG INTO THE AGAR WITH THE WIRE.
9. Flame the neck of the bottle and replace the cap.

10. Flame the wire and place it on the bench.

11. Repeat with the other bacterium, i.e. select a red colony at Stage 1 of Part 4.

12. Label your cultures with the name of the organism and the date.

The slope cultures can be left at room temperature for 2–3 days to ensure adequate growth. They should then be stored in a refrigerator.

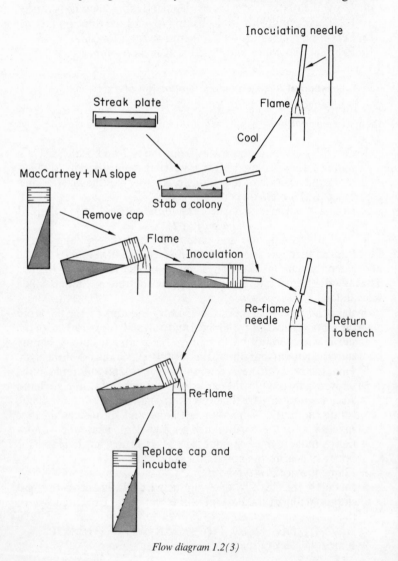

Flow diagram 1.2(3)

QUESTIONS

1. How effective were the two methods in terms of isolating the red and yellow bacteria?
2. Are there any differences between the streak and pour plates?
3. Can you explain how both methods effected separation of the bacteria?
4. Why is it essential to mix the bacteria and media by the methods used in the experiment and not by shaking up and down or inverting?
5. Why are loops and wires heated to red heat before and after use?
6. Why are the ends of Pasteur pipettes plugged with cotton wool?
7. Why should you be careful when using pipettes not to suck liquids up into the cotton wool or the teat?
8. Why are tubes or bottles held as near to the horizontal as possible when their caps have been removed?

EXPERIMENT 1.3

PREPARING STAINED PREPARATIONS OF BACTERIA

THEORY

Live bacteria are difficult to examine in ordinary wet preparations and so stained preparations are normally made. In this way various aspects of the bacterial cell can be examined more easily, for example size, shape, etc. Preparations can be made from bacterial colonies growing on plates or slopes, or from broth cultures or directly from a specific environment.

Bacteria are usually stained with basic aniline (coal tar) dyes as bacterial cytoplasm has a strong affinity for these stains. Thus methylene blue, crystal violet, or basic fuchsin are often used.

In this experiment you will examine natural yoghurt for the presence of bacteria using a simple staining technique.

PROCEDURE

Part 1. Preparation of a film of bacteria

1. Remove a grease-free slide from the jar of alcohol USING A PAIR OF FORCEPS. Allow any excess alcohol to run off the slide back into the jar. (NOTE. Grease-free slides have been prepared by soaking the slides in chromic acid, washing thoroughly in distilled water and storing in alcohol.)
2. Ignite the alcohol in a Bunsen flame and when the alcohol has

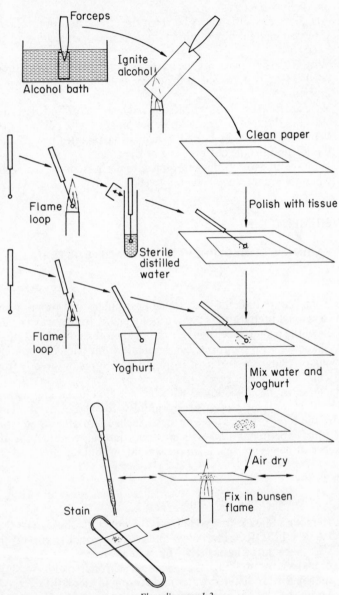

Flow diagram 1.3

finished burning, place the slide STILL WITHOUT FINGER-ING IT, on a clean sheet of paper on the bench.
3. Polish the slide with a tissue or a piece of paper towel. DO NOT TOUCH THE SURFACE OF THE SLIDE WITH YOUR FINGERS. Hold it by the edges at one end only.
4. Flame a loop and transfer a drop of sterile distilled water to the middle of the slide.
5. Re-sterilise the loop and use it to take a VERY SMALL amount of yoghurt from its container.
6. Transfer the yoghurt to the water on the slide and mix them thoroughly using the loop. Spread the mixture out over the slide to produce an even thin film (see *Flow Diagram 1.3*).
7. Allow the slide to air dry on the bench. Alternatively, the slide can be dried more rapidly by holding it WELL ABOVE a Bunsen flame. If the slide is heated too rapidly, however, the film will be ruined so heat very slowly and gently.
8. Label the slide with a distinguishing mark on the film side, using a wax pencil.

Part 2. Fixing the film

Having air dried your slide, pass it quickly, two or three times through the blue part of the Bunsen flame, holding the slide film-side up.
NOTE. THE SLIDE MUST NOT BECOME TOO HOT TO TOUCH. If it does, discard it and start again, i.e. at Stage 1.

The purpose of this stage is to heat-fix the film. This kills the bacterial cells, coagulates the cell proteins and causes the bacteria to adhere firmly to the slide. Thus they are not washed off the slide during the staining process.

Part 2. Staining the film

1. Place the slide film-side uppermost on a staining rack across the sink.
2. Add sufficient stain to cover the film, and leave it there for 2 min.
3. Rinse the stain off with tap water.
4. Remove the slide from the rack and allow excess water to drain off by holding the slide vertically on a tissue.
5. Carefully blot dry using clean blotting paper or filter paper.
6. Examine the film under the microscope (see Appendix to Experiment 1.3, page 26).
   (*a*) If you examine using the high-power objective ($\times$ 40), place a drop of Canada balsam on the film and cover with a cover slip.

(b) If you are using an oil-immersion objective, place a drop of immersion (Cedar wood) oil directly on the film. There is no need to mount with a cover slip.

7. Compare the stained film with one that has not been stained and then answer the following questions.

QUESTIONS

1. What are the shapes of the bacteria present?
2. How does their size compare with other cells you have seen?
3. Why is it essential to use grease-free slides?
4. Why should care be taken not to overheat films when drying or heat-fixing?

Appendix to Experiment 1.3

MICROSCOPY

Many of the organisms we study in biology are too small to be seen with the unaided eye, and this is also the case with parts of larger organisms which we may want to study.

The compound microscope is one of the most important instruments in the biology laboratory because it enables the biologist to make critical observations of very small objects. Always remember that the microscope is an expensive, delicate instrument, and damage due to improper handling can be expensive and may prevent the use of the microscope by other students.

The quality of the image, and thus the information you obtain with the microscope, depends on your ability to set up and use the microscope correctly. Setting the microscope up takes only a few minutes, but doing this properly is well worth while in terms of the results you will obtain.

PRACTICAL DIRECTIONS FOR USE OF THE MICROSCOPE

1. Familiarise yourself with the parts of a compound microscope (see *Figure 1.7*).
2. Position the microscope on the bench so that you are in a comfortable working position.
3. Rotate the nosepiece until the low power ($\times 10$) objective clicks into its working position in line with the draw tube (body tube).
4. Open the iris diaphragm fully.
5. If the microscope has a separate light source, position the lamp about 8 in from the microscope mirror. Check that the flat side

of the mirror is in use—the concave surface is only used when there is no substage condenser. Make sure that little or no light is shining on the top of the stage.

6. Remove the eyepiece and adjust the mirror position until the back lens of the objective is filled with light from the lamp. Return the eyepiece to the tube.

N.B. Omit Stage 6 if the microscope has an integral light source.

7. Clip a slide plus specimen on the microscope stage, and focus on it with the low-power lens. Lower the objective using the coarse adjustment and LOOKING FROM THE SIDE OF THE MICROSCOPE, until the objective is just a few millimetres from the specimen (cover slip). Note that some microscopes have a stop below which you cannot lower the low-power lens. If so, lower the lens to this stop.

Now carefully focus by looking down the microscope and turning the coarse adjustment back, i.e. raising the low-power lens.

If the specimen is very small (or motile), you may have difficulty in finding it. In this case obtain the correct focal plane by focusing on the edge of the cover slip or on a larger specimen.

8. FOCUSING THE CONDENSER

Hold the point of a pencil against the light source, i.e. against the light bulb, and looking down the microscope, adjust the position of the condenser until a sharp image of the pencil point is superimposed on the specimen.

N.B. The condenser is usually in focus when it is close to the underside of the slide on the stage.

9. ADJUSTING THE IRIS DIAPHRAGM

The iris diaphragm should already be open fully (Stage 4 above). The field of view may be hazy, especially when working at a high power. Close the iris diaphragm until the field just begins to darken and the image is sharp. (Alternative method. Remove the eyepiece and note the image of the iris diaphragm in the back lens of the objective. Close the diaphragm until the light area is about $\frac{2}{3} - \frac{3}{4}$ the total light area.) Note that this is the correct aperture for this particular lens only. For other objectives re-adjustment is necessary.

N.B. The iris diaphragm should not be used to control the light intensity. However, most student microscopes do not have a variable light intensity source so that you may have to use the iris diaphragm incorrectly, particularly when viewing living specimens.

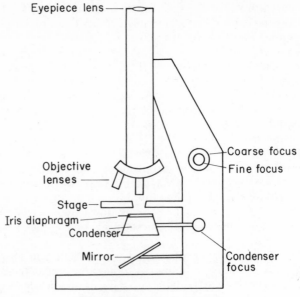

Eyepiece lens

Objective lenses

Coarse focus

Fine focus

Stage

Iris diaphragm

Condenser

Mirror

Condenser focus

Figure 1.7 *The optical microscope*

10. CHANGING OBJECTIVES

To use the high-power objective proceed as follows:
Ensure that the object is in the CENTRE of the low-power field of view. WATCHING FROM THE SIDE rotate the nosepiece to bring the high-power objective into position for use and make sure that (*a*) the lens does not hit the stage clips and (*b*), that it does not hit the cover slip. Now look down the microscope and you should be able to see your specimen even if it is out of focus. Bring it into focus using the FINE adjustment.

Many microscopes are arranged so that when the low-power lens is in focus you can change to the high-power lens without refocusing, or with minimal refocusing. Do not rely on this being the case, but always follow the procedure as above.

CAUTION. Never use the coarse adjustment when focusing the high-power objective as the working distance is so short that you may crush the slide and damage the lens.

11. USING OIL-IMMERSION OBJECTIVES

When using very high-power objectives (e.g. ×90 or ×100), it is necessary to place oil between the slide and the lens (see *Figure 1.9*). Carefully position the object to be examined in the centre of the high-power field and clip the slide firmly into position. Rotate the high-power lens out of the way. Place a small drop of immersion oil on top of the cover slip or smear so that this is directly above the condenser lens.

Raise the microscope body, i.e. rack back the coarse adjustment, and rotate the oil-immersion lens into a working position. Looking from the side, lower the lens carefully onto the oil drop. If the lens is the telescoping type, i.e. spring-loaded, continue lowering until the lens just begins to telescope. If not of this type, then just lower the lens until it is as close to the slide as possible. Now look down the microscope and raise the lens (rack back) until the object comes into focus.

Always remember that the working distance of the oil-immersion lens is very short (approximately 1·2 mm) and so great care is needed when it is in use.

12. SELECTING THE CORRECT OBJECTIVE

When using the microscope never think purely in terms of using the highest magnification available. Always select the correct lenses for the work you are doing.

Low-power objectives have a greater field depth than those of higher power and their use enables a greater view of preparations, for example with a thick object more levels are in focus at one time.

Always work upwards from low to higher-power objectives, NEVER the other way round.

13. CARE OF YOUR MICROSCOPE

Observe the following:
  (a) Never allow liquids to remain on the microscope stage.
  (b) Cover the microscope when not in use to keep out dust.
  (c) Clean the lenses with the specially provided lens tissue. Do not use handkerchiefs or laboratory coats, etc. These may scratch the lenses and cause irreparable damage.
  (d) Always support the microscope carefully when moving it.
  (e) Never dismantle lenses or other parts of the microscope. This is a SKILLED job and, if necessary, must be left to qualified personnel.
  (f) Always remove immersion oil immediately after use. Wipe off with a tissue.
  If you cannot obtain a clear picture, check:
    Incorrect setting up.
    Dirt on the lenses.
    Water on the cover slip surface.
    Congealed immersion oil (or even Canada balsam) on the lenses.

If the latter is the case, it can be removed using xylene but use very sparingly, as it loosens the lens mounting.

SOME THEORETICAL ASPECTS OF MICROSCOPY

Most modern microscopes have the power of their objective lenses engraved on the side of the lens. Thus:

low power $- \times 10$
high power $- \times 40$
oil immersion $- \times 100$

The eyepiece similarly has its magnification stamped into the top ring of the unit. Thus the total magnification can be easily calculated by multiplying the magnifying power of the eyepiece and objective together, for example for a $\times 10$ objective and a $\times 15$ eyepiece the total magnification $= 10 \times 15 = \times 150$.

An important property of the microscope, however, is not simply its ability to magnify, but its ability to separate objects which lie close together. This ability to show more detail clearly and so give more information is called the resolving power.

The unaided human eye can only separate two objects that are approximately 0·2 mm apart, at a distance from the eye of 25 cm. Objects closer than this appear as a single object. A good example of this is to look at a newsprint photograph. First look at it with the unaided eye. You will see the picture and you may be able to see dots. Now look at it with a hand lens and you will see that the picture is resolved into a large number of black and white dots.

The resolving power of the microscope is dependent upon a number of factors. These are:

(*a*) The wavelength of the light used.
(*b*) The correct iris diaphragm aperture.
(*c*) Correct condenser focusing.
(*d*) The light-gathering power of the objective lens.
(*e*) The observer's eyesight.
(*f*) Aberrations in the optical system.
(*g*) The contrast between the object and its background.

The most important consideration is usually the light-gathering power of the objective lens which is expressed in terms of its numerical aperture (N.A.).

$$N.A. = \mu \sin \theta$$

where μ is the refractive index of the medium between the lens and the object, and θ is the half angle of light entering the objective lens (θ is, of course, influenced by the working distance and the lens diameter). See *Figure 1.8.*

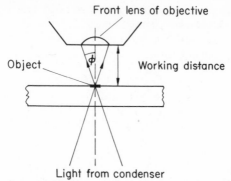

Figure 1.8 *Diagram illustrating the light path through to the objective lens*

Note that numerical apertures are normally engraved on the sides of the objectives, for example 1·2 for an oil-immersion ($\times$ 100) lens.

The resolving power of the objective lens is related to the numerical aperture and to the wavelength of the light used by the formula:

$$d = \frac{0·6\lambda}{\text{N.A.}}$$

where d = the smallest distance that can be separated (i.e. minimal distance between two resolvable objects),

λ = wavelength of light employed,

N.A. = numerical aperture of the objective.

For oil-immersion lenses with short focal length and giving high magnification, the light-gathering power of the lens and hence the N.A. can be increased by employing a medium between the lens and the object with a refractive index (R.I.) greater than air (R.I. = 1·0).

Usually this medium is cedar wood oil of approximately refractive index 1·5, which is almost the same R.I. as glass. For some lenses, water (R.I. = 1·3) can be used (see *Figure 1.9* below).

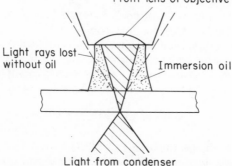

Figure 1.9 *To show oil immersion lens* in situ *with oil, and to show light pathways*

A top quality oil-immersion lens may have an N.A. of 1·4. If these oil-immersion lenses are used with green light (wavelength 550 nm), to which the human eye is most sensitive, then we can substitute these values in the equation:

$$d = \frac{0·6 \times \lambda}{\text{N.A.}}$$

$$= \frac{0·6 \times 550}{1·4} = 236 \text{ nm}$$

(Note: 1 μm = 10^{-3} mm
1 nm = 10^{-6} mm or 10^{-3} μm)

This value is approximately the maximum resolving power of the optical microscope. In order to achieve greater resolution other microscopes, for example the electron microscope, must be used.

Construct a table for your microscope showing the following data:

Eyepiece magnification	Objective magnification	Total magnification	N.A.	Resolving power

EXPERIMENT 1.4

METHODS OF STERILISATION

THEORY

The apparatus and media used in experimental work with micro-organisms must be sterile, i.e. free from living organisms. If they are not, then contaminating organisms may influence experimental results, preventing growth of the experimental organism, etc. The experiments in subsequent chapters necessitate sterilised media and apparatus and so in this experiment you will consider methods of sterilisation.

There are a wide variety of procedures available for sterilisation, namely:

1. Dry-heat sterilisation—hot air oven at 160°C for $1\frac{1}{2}$ h.
2. Boiling or steaming at atmospheric pressure.
3. Steaming under pressure—**autoclaving**. Using an autoclave or

pressure cooker, at either 10 lb/in² or 15 lb/in² pressure which produces temperatures of 115°C and 121°C respectively.
4. Ultraviolet light irradiation.
5. Ionising radiation.
6. Filtration.
7. Chemical methods.

The choice of method will depend upon the purpose of sterilisation and the nature of the material being sterilised. Thus in order to sterilise a plastic Petri dish, steam heating or autoclaving could not be used as the dish would melt; irradiation by gamma rays, however, can be employed.

In this experiment you will test the effectiveness of some of these methods of sterilisation. The material used will be soil, which contains a wide variety of micro-organisms, such as:

(a) Vegetative bacterial cells.
(b) Bacterial spores resistant to adverse conditions. e.g. heat or chemicals.
(c) Fungal mycelium.
(d) Fungal spores.

PROCEDURE

N.B. Throughout this experiment, use the aseptic techniques you have learnt in Experiments 1.1–1.3. Handle cotton wool plugs as if they were bacteriological tube caps (see Experiment 1.2, procedure stages Part 1, Stage 6 and *Figure 1.4*).

1. Weigh out 1 g of soil into each of two bacteriological tubes and plug them with cotton wool plugs.

Preparation of cotton wool plugs

Take a small square of non-absorbent cotton wool (approximately 4 cm square) and fold in the two sides so that they meet. Starting from the bottom roll up tightly (see *Figure 1.10*). A plug correctly made can be removed and replaced from a tube or flask several times without unrolling. For flask plugs simply use a larger square of cotton wool and follow the same method.

2. Label these two tubes, Expt 1.4, Group No. . . . , Tube 1 (or 2). Place both tubes into a hot-air oven at 160°C and leave them for 1½ h. (Continue with the experiment and see Stage 10 (later) for dealing with these tubes.)
3. Take a sterile 250 cm³ flask and prepare a cotton wool plug for the flask. Weigh out 8 g of soil and put it into the flask. Add 80 cm³ sterile distilled water to the flask and mix the soil and water thoroughly.

34

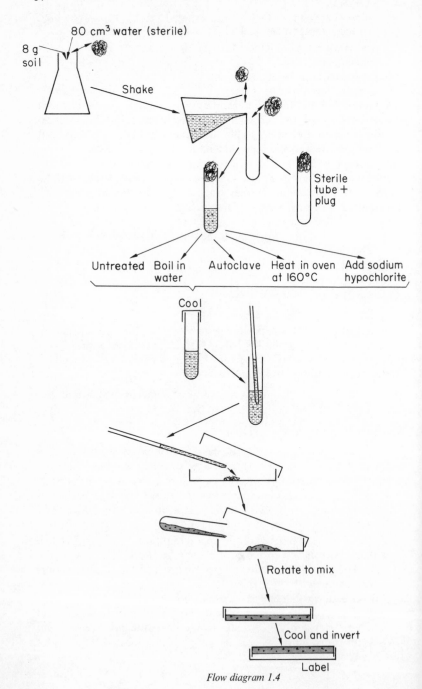

Flow diagram 1.4

4. Take eight bacteriological tubes and make cotton wool plugs for them. Label them using a wax pencil (or an indelible felt-tip pen) with Expt 1.4, your Group No., and the numbers 3–10 respectively.
5. Reshake the soil suspension and pour it into the eight test-tubes so that each test-tube is one-third full.
6. Leave Tubes 3 and 4 untouched.
7. Take Tubes 5 and 6 and put them in a bath of boiling water. Leave them for 10 min, then take them out and allow to cool (see also Procedure, Stage 20).
8. To each of tubes 7 and 8 add two drops of a 10% sodium hypochlorite solution.

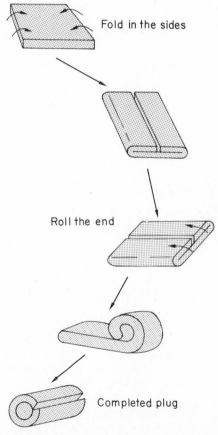

Fold in the sides

Roll the end

Completed plug

Figure 1.10 *Procedure for rolling a cotton wool plug*

9. Place tubes 9 and 10 in an autoclave or pressure cooker. Cover the cotton wool plugs with foil (or greaseproof paper). Autoclave the tubes at 15 lb/in^2 for 15 min (see Appendix on p. 37 for details). Remove the tubes from the autoclave and allow them to cool.

10. Take Tubes 1 and 2 out of the hot-air oven and allow them to cool in a rack on the bench. Take a sterile pipette and add 10 cm^3 of sterile distilled water to each of the two tubes. Shake them to mix the soil and water.

11. Place all the tubes together in a rack on your bench.

12. Take ten sterile Petri dishes and five sterile 1 cm^3 pipettes.
Label five of the dishes Expt No. 1.4, NA, and number the tubes 1, 3, 5, 7, and 9, respectively.
Label the other five dishes Expt No. 1.4, MA, and number the tubes 2, 4, 6, 8, and 10 respectively.

13. Take one of the sterile 1 cm^3 pipettes. Shake Tube 1 to disperse the soil, and suck up 1 cm^3 of the suspension. Pipette this into the Petri dish bearing the Tube 1 label. Repeat using the same pipette for Tube 2.
BE VERY CAREFUL NOT TO LET THE PIPETTE TOUCH THE BENCH OR ANYTHING ELSE until you have finished with it.

14. Now repeat Procedure 13 for each of the pairs of tubes, i.e. 3 and 4, 5 and 6, etc.

15. Obtain five tubes (bottles) of nutrient agar from the 45°C water-bath. Check to ensure they are molten by tilting.

16. Take the Petri dish bearing the label Tube 1, and pour in the nutrient agar. Mix the agar and soil suspension by placing the fingers of your right hand on the top of the Petri dish and, by moving your hand in a circular manner, rotate the dish first clockwise and then anti-clockwise. THIS ACTION MUST BE CARRIED OUT GENTLY otherwise you will agitate the agar so violently that it will splash the Petri dish lid.

17. Repeat Stage 16 for Petri dishes bearing the tube numbers 3 5, 7, and 9.

18. Obtain five tubes (bottles) of malt extract agar. Pour plates as described above in Stage 16, using the dishes bearing the label MA, and the tubes numbered 2, 4, 6, 8, and 10.

19. Leave all your dishes at room temperature for a few days.

20. Keep the tubes numbered 5 and 6.

21. After one day (i.e. the next day after you have set-up the experiment), again put Tubes 5 and 6 in a bath of boiling water for 10 min. Remove and allow to cool. Take 1 cm^3 samples and make two plates, one of NA and one of MA. Label them Expt

1.4, Group No. . . ., Tube 5, NA, set B and Expt 1.4, Group No. . . ., Tube 6, MA, set B.

22. After a few days examine all the plates. Make an estimate of the numbers of colonies appearing on each of the plates. (There is no need to count these very accurately.) Compare your results with those of other groups. Use your results to compare the effectiveness of the sterilisation methods used, and to answer the following questions:

QUESTIONS

1. What advantages are there in using an autoclave rather than a hot-air oven as a sterilising agent?
2. How effective is boiling water as a sterilising agent?
3. What is the effect of boiling after an interval of 24 h? Can you explain this result?
4. How effective is sodium hypochlorite as a sterilising agent?
5. When is it convenient to use a chemical agent for sterilisation in the laboratory?
6. Why is it important to use non-absorbent cotton wool for plugging tubes and flasks, rather than absorbent cotton wool? Why is it necessary to cover the cotton wool plugs of Tubes 9 and 10 when they are autoclaved?

Appendix to Experiment 1.4

AUTOCLAVING

As noted above, autoclaving means steam sterilisation under pressure. Autoclaves or pressure cookers are marketed in many different forms and the use of your particular model will be demonstrated by your teacher, however, they all have the basic similarities shown in *Figure 1.11*. These are:
1. A body which can withstand high pressures.
2. A lid with either a locking device or locking screws.
3. A sealing washer between lid and base.
4. A safety valve.
5. A pressure regulator or in a pressure cooker variable weights.
6. A pressure gauge may be present.

When using the autoclave make sure you remember the following points:
(a) Always ensure that the volume of water in the bottom is sufficient to generate enough steam. This is usually about one litre in a small autoclave.

(b) When loading the autoclave make sure there is sufficient space between apparatus to allow for the circulation of steam.

(c) Open the steam outlet (remove the weights if you are using a pressure cooker).

(d) Secure the lid in position (set the automatic pressure device).

(e) Light the gas or switch on the heating element.

(f) Allow the water in the autoclave to boil and generate steam. When a steady, i.e. constant, jet of steam is issuing, close the steam outlet (add the pressure cooker weights).
N.B. It is essential that all the air should be displaced, i.e. driven out by the steam, as only pure steam will produce the correct temperature at a particular pressure.

(g) Watch the gauge and when the required pressure has been reached, regulate the heat to give a constant pressure (if this is not done automatically).
The most commonly used pressures are from 10–15 lb/in^2, giving temperatures from 116–121°C. These are maintained from 10 to 15 min, the time starting when the required pressure is reached. In Experiment 1.4 we use 15 lb/in^2 for 15 min.

(h) After the time at correct pressure has elapsed, turn off the heat and allow the autoclave to cool. When the pressure gauge reads

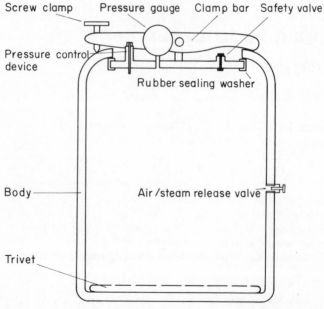

Figure 1.11 *An autoclave*

zero, open the steam outlet (remove the weights). **DO NOT OPEN THE STEAM OUTLET BEFORE THIS TIME,** otherwise the steam will rush out of the autoclave, the pressure will fall drastically and any liquid media may boil out of their containers.

(*i*) Allow the autoclave to cool even further before removing the lid. This is important since even though the pressure inside is atmospheric, liquids may not have had time to cool to 100°C (they may be **superheated**) and may boil vigorously if the containers are disturbed.

(*j*) Remove apparatus from the autoclave and allow to cool at room temperature. Agars go directly into a 45°C water-bath if required immediately.

(*k*) Leave the autoclave as tidy as you found it.

2
Nutrition

INTRODUCTION

The nutrition of organisms is basic to their survival and either they manufacture their own food from inorganic materials, i.e. are **auto-trophic**, or they utilise a variety of ready-made food substances (many of which are the products of autotrophs), i.e. are **heterotrophic**.

Many heterotrophs encounter foods in more complex forms than they can immediately utilise in their metabolism. This is because large macromolecules cannot be absorbed directly often due to size alone. The heterotrophs therefore must employ some method of simplifying these foods. Usually foods can be categorised into one of the three major food types, Fats, Carbohydrates, Proteins; or else they are organic compounds such as vitamins; or are inorganic salts, minerals or water. The major food types can all be broken down into simpler sub-units by living heterotrophs:

> Proteins into amino acids,
> Carbohydrates into monosaccharide sugars,
> Fats into fatty acids and glycerol.

These heterotrophs then, have the ability to use the macromolecules that are in the environment, and they do so by secreting enzymes (biological catalysts) largely extracellularly, changing complex foods into simpler, utilisable units.

The experiments in this chapter exemplify many of the above concepts. Thus the first two experiments examine the nutrient requirements of an autotroph and a heterotroph, whilst Experiments 2.3 and 2.4 show aspects of extracellular digestion. The final experiment in the chapter deals with the interesting effect a vitamin has on the growth of two heterotrophic organisms.

EXPERIMENT 2.1

NUTRIENT REQUIREMENTS IN AN AUTOTROPHIC ORGANISM—*CHLORELLA*

THEORY

Chlorella is a green alga, it has small green chlorophyll-containing cells and is classified as a member of the Chlorophyta. Each globular unicell of this organism has a single nucleus and large chloroplasts. It has been used widely for research into plant physiology.

Investigations using sand and water cultures with green plants, to investigate nutrient requirements, have been carried out since the 1850s, but suffer from many disadvantages. This experiment uses a smaller plant and can be regulated more easily.

PROCEDURE

1. Prepare twenty-two 250 cm³ conical flasks by cleaning thoroughly and washing with glass-distilled water.
2. Using the solutions of salts provided set up the following 11 flasks by reference to *Table 2.1*. Make a duplicate set of flasks.
   - *(1)* Complete medium made up to 25 cm³ with 0·1 % sodium bicarbonate solution.
   - *(2)* Complete medium made up to 25 cm³ with 0·1 % sodium bicarbonate solution.
   - *(3)* Complete medium made up to 25 cm³ with 1·0 % glucose solution.
   - *(4)* (*a*) Complete medium made up to 25 cm³ with boiled distilled water.
   - *(4)* (*b*) 25 cm³ of 0·1 % sodium bicarbonate solution.
   - *(5)* Complete medium, less nitrogen, made up to 25 cm³ with 0·1 % sodium bicarbonate solution.
   - *(6)* Complete medium, less phosphate, made up to 25 cm³ with 0·1 % sodium bicarbonate solution.
   - *(7)* Complete medium, less potassium, made up to 25 cm³ with 0·1 % sodium bicarbonate solution.
   - *(8)* Complete medium, less magnesium, made up to 25 cm³ with 0·1 % sodium bicarbonate solution.
   - *(9)* Complete medium, less iron, made up to 25 cm³ with 0·1 % sodium bicarbonate solution.
   - *(10)* Complete medium, less sulphur, made up to 25 cm³ with 0·1 % sodium bicarbonate solution.
3. Autoclave flasks No. 3 at 10 lb/in² for 15 min and allow to stand and cool.

Table 2.1 AMOUNTS OF NUTRIENT SALT SOLUTIONS FOR FLASKS 1–11

0·5 M solutions of salts	Media in flasks—amounts in cm³						
	Complete	Less N	Less P	Less K	Less Mg	Less Fe	Less S
KNO_3	5	0	5	0	5	5	5
K_2HPO_4	2	2	0	0	2	2	2
KCl	0	5	4	0	0	0	0
$MgSO_4$	2	2	2	2	0	2	0
$MgCl_2.6H_2O$	0	0	0	0	0	0	2
$NaNO_3$	0	0	0	5	0	0	0
Na_2SO_4	0	0	0	0	2	0	0
Na_2HPO_4	0	0	0	2	0	0	0
$FeSO_4.7H_4O$	1	1	1	1	1	0	0
$FeCl_3.6H_2O$	0	0	0	0	0	0	1

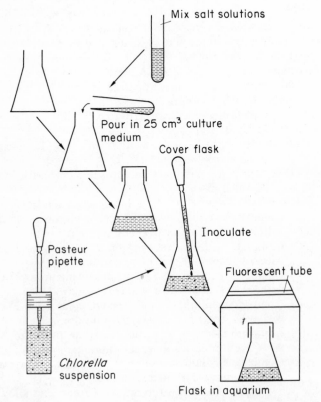

Mix salt solutions

Pour in 25 cm³ culture medium

Cover flask

Inoculate

Pasteur pipette

Fluorescent tube

Chlorella suspension

Flask in aquarium

Flow diagram 2.1

4. Take a suspension of *Chlorella* sp. in sterile distilled water and shake it to disperse the cells evenly.
5. With a Pasteur pipette (see Chapter 1), take up some of the *Chlorella* suspension.
6. Add two drops of the suspension to each of the flasks.
7. Cover each flask with a 100 cm³ beaker (or crystallising dish or polythene).
8. Place flasks *(1)* and *(3)* in the dark and leave them for 2 weeks.
9. Place the remaining flasks under a suitable light source. A good source is a 'Growlux' fluorescent tube, which emits more red light than standard fluorescent tubes.

ASSESSMENT OF RESULTS

10. Using a Pasteur pipette place one drop from the first flask on a clean glass slide. Examine it under the high-power lens of a microscope. Count the number of individual cells within the high-power field. If clumps are present, then count the number of individual cells in the clump. Do not select the field of view, simply move the slide whilst looking from the side of the microscope.
    Repeat this for ten fields of view.
11. Repeat Stage 10, for each of the 22 flasks. Make sure each flask has *its own* specific pipette.
12. Collect all your results and calculate an average for each of the 11 flasks.
13. Collect the results from the other groups in your class and tabulate them.
14. Draw a bar diagram of the average numbers of individuals against the eleven varying situations.

QUESTIONS

1. What is the purpose of putting sodium bicarbonate solution into the flasks?
2. What is the purpose of putting the flasks containing the glucose in the dark?
3. Why is flask (4a) set up with BOILED distilled water and not just ordinary distilled water, and what does flask (4a) show? What does flask (4b) show?
4. Can you account for the variation in the amount of growth that has occurred in flasks (1)–(10). Explain in terms of the physiology of *Chlorella*, the role of the missing nutrients in the flasks.
5. Why have we set up a duplicate set of flasks?
6. Why, when assessing numbers of individuals, must we take the

precaution of moving the slide when looking from the side of the microscope?

7. Why do we collect results from all the groups in the class?
8. Give reasons why you think this experiment is (a) better, or (b) as good as experiments using sand or water cultures to investigate nutrient requirements?

EXPERIMENT 2.2

NUTRIENT REQUIREMENTS IN A HETEROTROPHIC ORGANISM—*ASPERGILLUS NIGER*

THEORY

Aspergillus niger is a black mould belonging to the group Ascomycetes. It is a widely distributed organism occurring from the tropics to the arctic and can utilise a large variety of food substances.

This experiment, like Experiment 2.1 can be carefully controlled so that the precise nutrient requirements of this organism can be investigated.

PROCEDURE

1. Take twenty 250 cm³ conical flasks. Thoroughly clean them and complete the cleaning by washing several times with distilled water.
2. Set up the flasks in the following way.
   By reference to *Table 2.2* prepare the following sets of solutions:
   (1) Complete medium made up to 25 cm³ with distilled water.
   (2) Complete medium made up to 25 cm³ with 1% glycerol solution.

Table 2.2 TO SHOW AMOUNTS OF SALT SOLUTIONS IN FLASKS 1–10

0·5 Molar solutions of salts	Media in flasks—amounts in cm³						
	Complete	Less N	Less P	Less K	Less Mg	Less Fe	Less S
KNO_3	5	0	5	0	5	5	5
K_2HPO_4	2	2	0	0	2	2	2
KCl	0	5	4	0	0	0	0
$MgSO_4$	2	2	2	2	0	2	0
$MgCl_2.6H_2O$	0	0	0	0	0	0	2
$NaNO_3$	0	0	0	5	0	0	0
Na_2SO_4	0	0	0	0	2	0	0
Na_2HPO_4	0	0	0	2	0	0	0
$FeSO_4.7H_2O$	1	1	1	1	1	0	0
$FeCl_3.6H_2O$	0	0	0	0	0	0	1

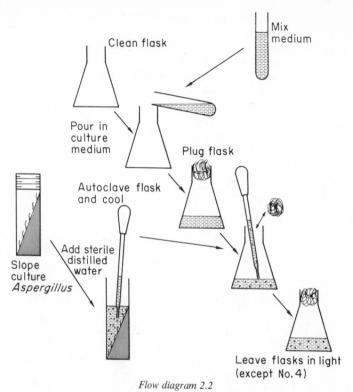

Flow diagram 2.2

(3) Complete medium made up to 25 cm³ with 1 % glucose solution.

(4) Complete medium made up to 25 cm³ with 0·1 % sodium bicarbonate solution.

(5) Complete medium, less nitrogen, made up to 25 cm³ with 1 % glucose solution.

(6) Complete medium, less phosphate, made up to 25 cm³ with 1 % glucose solution.

(7) Complete medium, less potassium, made up to 25 cm³ with 1 % glucose solution.

(8) Complete medium, less magnesium, made up to 25 cm³ with 1 % glucose solution.

(9) Complete medium, less iron, made up to 25 cm³ with 1 % glucose solution.

(10) Complete medium, less sulphur, made up to 25 cm³ with 1 % glucose solution.

Make up a duplicate set of flasks.

Label all the flasks carefully to show the contents.

3. Plug the flasks with a cotton wool plug.
4. Autoclave all the flasks except No. 4 at 10 lb/in^2 for 15 min. Allow them to stand and cool.
5. **Inoculation**. Take an agar slope which has a good growth of *Aspergillus* on it, and which shows the presence of black spores. Add to the slope a small amount of sterile distilled water. Gently shake the culture bottle to produce a spore suspension.
6. Using a sterile Pasteur pipette, take up an amount of this spore suspension. Add two drops of the suspension to each of the 20 flasks.
7. Place flasks No. 4 in the light.
8. Place all the remaining flasks in the dark. Leave them for 1 week.

Results

9. Devise a scale, for example 0–10, to assess the growth in the flasks. Make this assessment on a purely visual basis and use such factors as amount of mycelium, numbers of black spores or any other suitable factor.
10. In order to obtain a more quantitative assessment proceed as follows:
    (*a*) Weigh a coarse glass sintered crucible. Record the weight. Label the crucible with the flask number. Set the crucible in a holder on the top of a Buchner flask as in *Figure 2.2* on page 55.
    (*b*) Pour the contents of a flask into the crucible, and extract the liquid by turning on the water pump.
    (*c*) If necessary wash out the flask to remove all the mycelium or remove with the aid of a spatula.
    (*d*) Repeat stages (*a*)–(*c*) for all 20 flasks.
    (*e*) Dry the crucibles in an oven at 105°C.
    (*f*) Take the crucibles out of the oven and place them immediately in a desiccator and leave them to cool.
    (*g*) Re-weigh all the crucibles to a constant weight.
11. Calculate an average dry weight for each of the 10 flask situations (1)–(10). If possible, collect the results from other groups in the class and make up a composite average.
12. Draw a bar diagram of the dry weight in the different nutrient situations.

QUESTIONS

1. Can *Aspergillus* grow without an organic carbon source? Can it grow with just sodium bicarbonate in the light?

2. Is glycerol (a 3 carbon compound) as valuable a carbon source as glucose (a 6 carbon compound), for growth in *Aspergillus*?
3. Is there any solution that does not provide enough nutrients for any growth? Can you explain this?
4. Are there alternative food sources as good as glucose for this organism?
5. In terms of the physiology of *Aspergillus*, account for the variation in growth in flasks (1)–(10). Explain the role of the missing nutrients.
6. Does growth in a medium which lacks a nutrient mean that this nutrient is not essential for growth?

EXPERIMENT 2.3

PROTEIN DIGESTION IN BACTERIA

THEORY

Two different bacteria, *Serratia marcescens*, a red bacterium, and *Aerobacter aerogenes* occur quite commonly. *Serratia marcescens* is found in soil and *Aerobacter aerogenes* in association with rotting vegetation. They are both non-pathogenic and are heterotrophs.

Gelatin dissolved in water is a colloidal protein which at room temperature is stable and in the gel form. The protein is the disperse phase of the colloid. In the gel form the molecules of protein are held together by hydrogen bonds trapping water molecules in between the lattice so formed, see *Figure 2.1*. In the sol form the hydrogen bonds are broken.

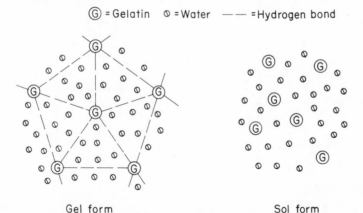

Ⓖ = Gelatin   Ⓞ = Water   —— = Hydrogen bond

Gel form   Sol form

Figure 2.1 *To show diagrammatically, the structure of gelatin*

PROCEDURE

1. Prepare a medium containing 12 g gelatin and 1 g of commercial Bovril per 100 cm³ of distilled water.
2. Take four bacteriological tubes and pour 5 cm³ of the medium into each tube.
3. Cover each tube with an Oxoid cap (or with aluminium foil), and autoclave for 15 min at 10 lb/in².
4. Remove the tubes from the autoclave, allow to cool and the medium to solidify.
5. Take a stab wire (see *Figure 1.1 (a)* on page 12).
   Flame-sterilise the stab wire by holding it almost vertically in the Bunsen flame, see *Flow Diagram 2.3*, and allow it to cool. Rub the stab wire on the surface of a slope bearing colonies of red *Serratia marcescens*.

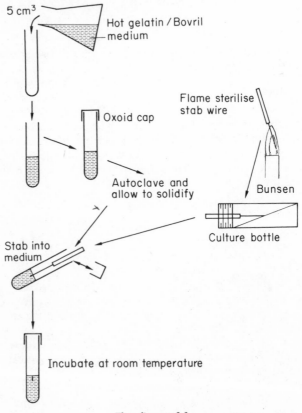

Flow diagram 2.3

REMEMBER TO FLAME THE MOUTH OF THE CULTURE BOTTLE BEFORE REPLACING THE CAP. Stab the wire into the centre of the first of your sterilised tubes. Flame the tube mouth and replace the Oxoid cap (foil cover). Label this tube No. 1.

6. Repeat Stage 5 but using *Aerobacter aerogenes* and label the tube No. 2.
7. Mark the remaining Tubes 3 and 4 as controls.
8. Leave to stand at room temperature for 3–5 days.
9. Examine the tubes, record your observations and then answer the following questions:

QUESTIONS

1. What does the liquefaction of the medium in Tube 1 represent?
2. Why has Tube 2 not shown the same liquefaction?
3. How do you think that the liquefaction is brought about? (i.e. how does the bacterium do it?)
4. What then is the difference between *Serratia marcescens* and *Aerobacter aerogenes* as heterotrophs?
5. Do you think either of the bacteria has any advantage over the other?
6. What do you think the liquid layer in Tube 1 consists of?
7. Why is there Bovril in the medium?
8. What is the purpose of Tubes 3 and 4 (controls)?

EXPERIMENT 2.4

STARCH DIGESTION IN *ASPERGILLUS NIGER*

THEORY

Aspergillus niger, as already noted in Experiment 2.1, is a black mould showing world-wide distribution. It is a heterotrophic organism which shows itself able to utilise a large variety of food substances, and in this experiment we will examine its ability to utilise starch.

Starch is a long-chain polysaccharide consisting of as many as 1000 glucose units. Its wide occurrence in plant material makes it an important substrate for the growth of micro-organisms.

The medium used to grow *A. niger* is buffered at pH 5·0. It contains all the necessary inorganic salts and agar. One gramme of starch per 100 cm^3 has been added as the only source of organic material.

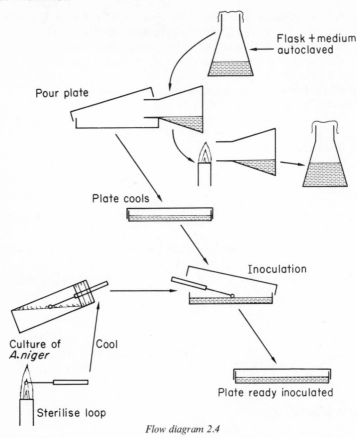

Flask + medium autoclaved

Pour plate

Plate cools

Inoculation

Culture of
A.niger

Cool

Sterilise loop

Plate ready inoculated

Flow diagram 2.4

PROCEDURE

The medium plus starch provided is sterile having been autoclaved at 10 lb/in² for 15 min.

1. Take four of the sterile glass Petri dishes.
2. Remove the foil cap (cotton wool plug) from the flask containing the medium.
3. Pour about 15 cm³ of medium into each of your four Petri dishes. Flame the mouth of the flask before replacing the foil cap. Remember; DO NOT lift the lid of the Petri dish more than is necessary.
4. Allow the dishes to cool and solidify. Label with the experiment number and then label the dishes 1, 2, 3 and 4.
5. Take an inoculating loop, and sterilise by heating it to red heat in a Bunsen flame.

6. Allow the loop to cool and then run it over the surface of a culture of *Aspergillus niger*, to pick up some spores.
7. Dab the loop on the centre of the medium (agar) in dish 1. Mark the spot on the bottom of the dish.
8. Repeat Stages 5 and 6 with dish 2.
9. Leave dishes 3 and 4 as controls.
10. Place all four dishes at room temperature in the laboratory and leave them for 5–7 days.

TESTING THE PLATES

11. By referring to the mark made when inoculating the plates, observe the extent to which the fungus has grown.
12. Draw exactly on a sheet of graph paper the area covered by the fungal mycelium in both dishes 1 and 2.
13. Flood dishes 1 and 3 with alcoholic iodine solution. Allow to stand for 2–3 min and then pour off the excess solution.
14. Flood dishes 2 and 4 with Benedict's solution. Again leave to stand for 2–3 min and pour off the excess solution. Autoclave these two dishes by bringing them to 5 lb/in² pressure and then immediately cooling.

INTERPRETATION OF RESULTS

15. Draw on the graph paper used in Instruction 12 the areas now coloured in all dishes. By comparing these with the results of the extent of fungal growth, answer the following questions.

QUESTIONS

1. Which carbohydrates give: (*a*) a blue-black colour with iodine? (*b*) no colour with iodine?
2. Which carbohydrates give green, yellow, or red precipitates with Benedict's solution?
3. Where on dish 1, has the iodine produced (*a*) coloured areas, (*b*) no reaction? Explain why this is.
4. Where on dish 2 has the treatment with Benedict's solution produced:
   (*a*) No reaction (i.e. blue areas)?
   (*b*) Green areas?
   (*c*) Yellow areas?
   (*d*) Red areas?
Explain any results obtained.
5. Can you relate the results in dish 1 in any way with the results obtained in dish 3?
6. Why is dish 3 treated with iodine, and dish 4 with Benedict's solution, and what does this show?

EXPERIMENT 2.5

THE EFFECT OF THE VITAMIN THIAMINE ON THE GROWTH OF HETEROTROPHIC ORGANISMS

THEORY

Nutritional requirements as described in the introduction include organic growth factors, required in small amounts, which are called vitamins.

This investigation involves the effect of the vitamin thiamine (B_1) on the growth of two fungi, *Phycomyces blakesleeanus* and *Aspergillus species*. The latter Ascomycete fungus was used in Experiments 2.2 and 2.4. *Phycomyces blakesleeanus* is a mould but it belongs to the group Phycomycetes.

PROCEDURE

The medium provided is a liquid medium containing all the necessary inorganic salts, and glucose, and is buffered at pH 5·0. It also includes asparagine, a nitrogenous compound related to the amino acid aspartic acid.

The vitamin solution contains 1 microgramme (µg) per cm^3.

1. Make up the following sets of 250 cm^3 conical flasks.
   Take twelve 250 cm^3 conical flasks and label them 1–12.
   Pour 50 cm^3 of medium into each flask.
   Add thiamine solution in the following amounts, to the flasks.

Table 2.3

Flask number	Thiamine solution cm^3
1 and 2	0·0
3 and 4	0·05
5 and 6	0·25
7 and 8	1·0
9 and 10	2·0
11 and 12	4·0

Plug the flasks with cotton wool plugs (see Chapter 1) and autoclave at 10 lb/in² for 15 min. Remove the flasks from the autoclave and allow them to cool.

2. Take the slope on which *Phycomyces blakesleeanus* is cultured and pour in about 10 cm^3 of sterile distilled water from a MacCartney bottle. Replace the cap of the culture tube and shake gently. This produces a spore suspension.

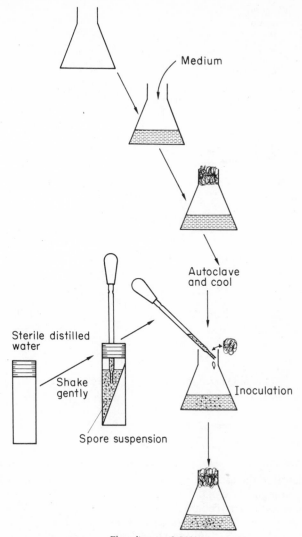

Flow diagram 2.5(1)

3. Using a sterile Pasteur pipette (see *Flow Diagram 2.5(1)* and Chapter 1) take up some spore suspension from the culture bottle.
4. Add two drops of spore suspension to each flask.
5. Repeat instructions 1–4 with a culture of *Aspergillus species*.
6. Leave both sets of flasks for 2 weeks at room temperature.

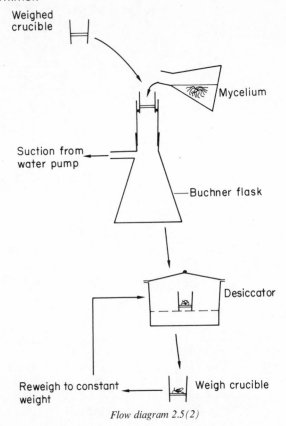

Flow diagram 2.5(2)

RECORDING YOUR RESULTS

7. Describe any growth that has occurred for both organisms, i.e. describe the amount of growth and the presence or absence of sporangia, or conidia.
8. Harvest the mycelium as follows:
   (*a*) Weigh a coarse sintered glass crucible. Record the weight. Label the crucible with the flask number. Set the crucible in a holder on top of a Buchner flask as in *Figure 2.5 (2)*.
   (*b*) Pour the contents of a flask into the crucible, and extract the liquid by turning on the water pump.
   (*c*) If necessary, wash out the flask to remove all mycelium or remove with a spatula.
   (*d*) Repeat Stages (*a*) to (*c*) for all 12 flasks.
   (*e*) Dry the crucibles at 105°C in an oven.

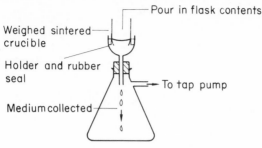

Figure 2.2

(*f*) Remove the crucibles from the oven and allow them to cool in a desiccator.

(*g*) Re-weigh all the crucibles to constant weight.

9. By collecting the results from other groups of experimentors, calculate an average dry weight of mycelium for each thiamine concentration.

10. Draw a graph of dry weight against thiamine concentration for each organism.

QUESTIONS

1. Is an external supply of thiamine essential for the growth of *Phycomyces* and *Aspergillus*?
2. If you consider that thiamine is not essential in the medium for one or other of these organisms, how does the organism grow without it? (Assume that thiamine is needed by all organisms.)
3. Why is the vitamin added in such small quantities, i.e. in $\mu g/cm^3$?
4. Does the increase in thiamine concentration have any effect on the amount of growth in *Phycomyces*?
5. How could you use your results to determine the amount of thiamine in a natural product, for example malt extract?
6. Why do you run duplicate sets of flasks and why do you collect results from other workers in order to arrive at an average?

3
Structure and life cycles

INTRODUCTION

In order to understand an organism completely, one must have knowledge of both its structure and life history. With many micro-organisms that affect man, either beneficially or harmfully, an exact knowledge of the life history and structure of the organisms is essential if either control is to be exercised, or the organisms are to be utilised to best advantage.

In this chapter, Experiment 3.4 examines yeast, an organism produced on a very large scale throughout the world for baking, brewing, as a source of biochemicals, and as a foodstuff. Other experiments investigate other micro-organisms of benefit to man. i.e. 3.2 – *Chlorella*, 3.6 – *Penicillium*, 3.7 – mushrooms. Although Experiment 3.1 does not investigate pathogenic bacteria, nevertheless the knowledge gained, and the techniques used could be applied when investigating pathogens.

In many experiments in other chapters of this manual you will use the organisms described in this chapter. A knowledge of their structure will be essential to the complete understanding of the concepts demonstrated by these later experiments.

EXPERIMENT 3.1

THE STRUCTURE OF BACTERIA

THEORY

The small size of bacteria makes examination of wet preparations under the optical microscope somewhat difficult. As mentioned in Experiment 1.3 therefore, bacteria are usually examined in fixed and stained preparations, which enable the shape and structure of the bacterial cells to be examined more easily. Also by staining cells taxonomic differences vital for identification can be revealed.

The most important staining technique used in routine bacteriollogy is **Gram's stain**. This was developed in 1884 by Christian Gram, a Danish physician, and remains a relatively simple staining method which is used to distinguish between two major groups of bacteria— the Gram-negative (—ve), bacteria, and the Gram-positive (+ve) bacteria.

Although originally this technique simply represented an empirical way of separating two types of bacteria the two groups do in fact show a remarkable number of other distinct features, for example:

(a) All spore-bearing bacteria are Gram +ve.
(b) Bacteria with polar (at the ends of the cells) flagella, are Gram —ve.
(c) Penicillin only inhibits the growth of Gram +ve organisms.
(d) There are chemical differences between the cell wall of Gram —ve and the cell wall of Gram +ve organisms.

Thus the Gram reaction has become a very useful taxonomic feature.

The Gram stain is a differential staining technique, i.e. it distinguishes between two groups of organisms, as compared with a simple staining technique, for example methylene blue, which stains all bacterial cells a uniform blue colour. In the Gram stain, heat-fixed films of bacteria are stained with crystal violet (or a related dye), and are then treated with iodine solution. This forms a crystal violet–iodine complex. The films are then flooded with an organic solvent, for example alcohol, and the cells of Gram —ve bacteria lose the crystal violet stain, whereas the cells of Gram +ve bacteria retain the purple dye. Normally the slides are then treated with a red dye so that in the finished smear Gram +ve cells appear violet, and Gram —ve cells appear red.

PROCEDURE

Part 1

Refer to Experiment 1.3, Part 1, on page 23 for the preparation of a film of bacteria.

1. Using the above technique prepare films of the following bacteria, using the 12–18 h broth cultures provided:

Bacillus megaterium   *Serratia marcescens*
Streptococcus lactis   *Vibrio* sp.
Staphylococcus epidermidis   *Spirillum serpens*
Sarcina lutea

N.B. It is essential that the technique is carried out with a young culture since some bacteria lose the Gram reaction when they are not actively growing.

2. Heat-fix the film. See Experiment 1.3, Part 2, page 25.
3. Carry out the Gram staining reaction on all films, as follows:
   (*a*) Cover the film with crystal violet stain. Leave for 1 min.
   (*b*) Wash the slide with tap water.
   (*c*) Rinse the slide with iodine solution. Then cover the film with iodine solution and leave for 1 min.
   (*d*) Wash very quickly in tap water—5 s.
   (*e*) Take up some 95% alcohol in a Pasteur pipette. Hold the slide at about 45 degrees and allow alcohol to drip on the upper edge of the film and run off the slide (see *Figure 3.1*). Continue until dye is no longer removed from the film.

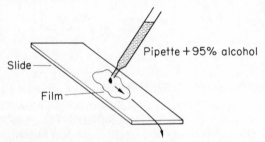

Figure 3.1 *To show Stage 3 (c) of gram staining reaction*

   (*f*) Immediately the dye stops coming out of the film, wash off any remaining alcohol with tap water.
   (*g*) Cover the film with 0·5% aqueous safranin and counterstain for 1 min.
   (*h*) Wash off the safranin with tap water and carefully blot dry. Label the slide with the name of the organism.
   Examine under the high-power oil-immersion lens of your microscope.
   N.B. Stages (*e*), (*f*), and (*g*), are critical. Too long in contact with aqueous safranin may remove stain from the Gram +ve organisms and give Gram —ve results.
4. Construct a table as shown below and enter your results.

Name of organism	Description of shape	Gram reaction	Dimensions

If an eyepiece micrometer and a stage micrometer are available, measure the dimensions of the bacteria and enter results in your table. Start with *B. megaterium* as it is the largest (see Appendix on page 60).

Part 2. Checking the isolation of bacteria

5. Take the slope cultures of *Serratia marcescens* and *Sarcina lutea* you produced in Experiment 1.2.
6. Obtain two bacteriological tubes and caps containing 10 cm³ of sterile nutrient broth.
7. Flame sterilise an inoculating loop and allow to cool.
8. Using aseptic technique, pick up a small loopful of the bacteria from the *S. lutea* slope culture. Transfer them to one of the nutrient broth tubes. Shake the tube and label it with your group number and the name of the organism.
9. Repeat Stage 8 using *S. marcescens*.
10. Incubate at room temperature for 12–18 h.
11. Prepare and stain films as above, Stages 1–4.
12. Examine and check on the success of your isolation experiment by comparing your isolates with films of pure cultures.

Part 3. Examining bacteria for endospores

Bacteria belonging to the genera *Bacillus* and *Clostridium* (both rod shaped) produce structures called **endospores**. These are produced inside the 'mother' cell and are highly resistant, remaining viable under conditions of heat, desiccation, and chemical treatment which would kill vegetative cells. The spores have a tough outer wall and dense cytoplasm containing little water. This makes these spores difficult to stain with simple stains like crystal violet and so special methods are used.

PROCEDURE

1. Using techniques from Part 1, prepare a heat-fixed film of the 48–72 h slope culture of *Bacillus cereus* (or *B. stereothermophilus*) provided.
2. Cover the film with 5% aqueous malachite green.
3. Twist some cotton wool on the end of a glass rod and dip it into alcohol. Ignite it and use it to gently heat the slide for 2 min. THE FILM SHOULD STEAM but DO NOT ALLOW THE STAIN TO BOIL.
4. Drain off the stain and wash in running water.
5. Counterstain with 0·5% aqueous safranin for 30 s.
6. Wash with tap water.
7. Blot carefully and examine under the oil-immersion lens. Spores stain green and vegetative cells stain red.

8. Sketch some spores to show their relative position and size within the vegetative cell.

N.B. If cultures were retained from Experiment 1.4, check these for endospores.

EXAMINING BACTERIA FOR MOTILITY USING THE 'HANGING DROP' METHOD

Many bacteria are capable of active movement by means of bacterial flagella, but these are too small to be seen under the optical microscope. Motility however, as indirect evidence of the presence of flagella, can be seen by suspending the bacteria in a hanging drop.

1. Place some Vaseline round the well of a cavity slide (see *Flow Diagram 3.1*).
2. Place a cover slip on a piece of clean paper on your bench. Obtain a sterile Pasteur pipette and take up some of the overnight broth suspension of *Serratia marcescens*.
3. Transfer a drop of this suspension to the centre of the cover slip.
4. Invert the cavity slide over the cover slip so that the well coincides with the drop of bacteria (see *Flow Diagram 3.1*). Press down LIGHTLY on the Vaseline ring.
5. Turn the slide quickly the right way up so that the drop of culture hangs from the cover slip into the well of the cavity slide.
6. Focus, using the low-power lens, on the edge of the hanging drop. Change to high power and examine for motility.
7. Repeat Stages 1–6 inclusive, for *Sarcina lutea*. (One of these bacteria is motile and the other is non-motile.)

Note: Be careful to distinguish between true motility, in which bacteria dart across the field, and Brownian movement, i.e. random vibration of bacteria due to molecular bombardment.

8. Having decided that you can recognise true motility, examine the other bacteria used in Part 1, and tabulate your results together with those already obtained.

Appendix to Experiment 3.1

THE CALIBRATION OF AN EYEPIECE MICROMETER USING A STAGE MICROMETER

Refer to *Figure 3.2* on page 62 showing a microscope eyepice. The image produced by the objective lens system is collected by the field lens of the eyepiece and focused in the plane of the field stop. You then view it by using the eye lens.

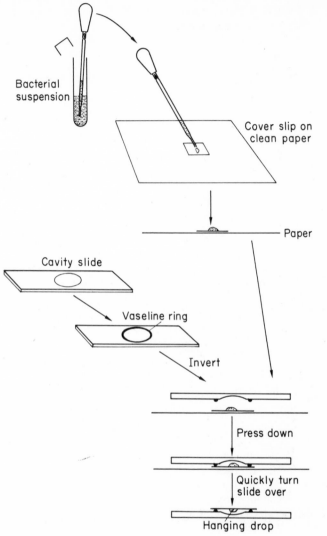

Bacterial suspension

Cover slip on clean paper

Paper

Cavity slide

Vaseline ring

Invert

Press down

Quickly turn slide over

Hanging drop

Flow diagram 3.1

When the eyepiece micrometer is in position, it is in almost the same position as the image and so when you look down the microscope you see an image and superimposed on this, the scale of the eyepiece micrometer. Since, however, the scale is mounted in between two glass discs, or etched on the surface of a disc, it may not coincide absolutely with the image, i.e. it will be very slightly out of focus.

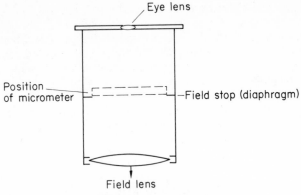

Figure 3.2 *An eyepiece*

PROCEDURE

1. Obtain an eyepiece micrometer (see *Figure 3.3*).
2. Unscrew the eye lens from the eyepiece and carefully drop the eyepiece micrometer on the field stop (see *Figure 3.2*).
   N.B. If the scale on the eyepiece is etched on the surface, make sure the scale side is face downwards.
3. Replace the eye lens and put the eyepiece back into the draw tube.
4. Set up the microscope as normal.
5. Now obtain a stage micrometer (see *Figure 3.3*).
6. Place the stage micrometer on the microscope stage, and focus on the scale with the low power objective. Decide what each scale unit is in millimetres.
   (*Note:* commonly the scale is 1 mm in length divided into one hundred parts, i.e. each unit is 0·010 mm or 10 µm (micrometres).)
7. Rotate the high power objective into position and move the stage micrometer so that the scale is central in the field of view. Secure it with the stage clips.
8. Place a drop of immersion oil on the slide and focus with the oil-immersion lens.
   TAKE GREAT CARE AS THESE SLIDES ARE EXPENSIVE.

Eyepiece micrometer   Stage micrometer

Figure 3.3 *Stage and eyepiece micrometers*

9. Rotate the eyepiece until the scales are parallel and partly superimposed.
10. Count the number of eyepiece divisions that are equivalent to a whole number, for example 0·010 mm, on the stage micrometer.
11. From this figure calculate the number of micrometres equivalent to one large eyepiece division, using oil immersion.
12. Now using the stained bacterial films, measure the size in eyepiece units of each bacterial cell. Measure lengths and breadths. Convert these figures to sizes in micrometres.

EXPERIMENT 3.2

THE LIFE CYCLE OF A GREEN ALGA—*CHLORELLA*

Chlorella is a unicellular, non-motile alga, the nutrient requirements of which are investigated in Experiment 2.1. This practical experiment is an investigation into the life cycle of *Chlorella* as an example of a green unicellular alga.

PROCEDURE

1. Mount a small drop of the culture of *Chlorella* on a clean glass slide. Use a clean pipette to remove the drop from the culture. Cover with a cover slip and examine under the high-power objective of the microscope, or preferably under an oil-immersion objective.
2. Refer to *Figure 3.5(a)* and acquaint yourself with the structures likely to be encountered in the cell. Refer also to *Figure 3.5(b)*. By carefully focusing with the fine adjustment locate the cell wall, chloroplast, and a more or less rounded body. *Note*. This is not the nucleus, but a structure called the pyrenoid. Draw a cell.
3. Search on your slide for stages (structures) containing two, four or eight cells. Also look for fragments of wall. Suggest a method of asexual reproduction (see *Figures 3.5(c–d)*). Draw as many different stages as possible.
4. Use the following technique to irrigate the culture with iodine in potassium iodide solution. Place a drop of iodine next to one side of the cover slip as in *Figure 3.4*. Touch the iodine with a needle (or brush) so that it is in contact with the liquid under

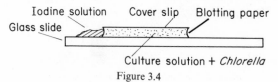

Figure 3.4

the cover slip and at the same time apply a small sheet of blotting paper to the opposite side of the cover slip, so drawing the iodine solution through under the cover slip. DO NOT REMOVE TOO MUCH SOLUTION.
5. Examine the stained slide. What changes have occurred? Draw the cell to show these changes. What do the changes suggest with regard to the function of the pyrenoid?
6. Make an illustrated diagram of the life cycle of this organism.

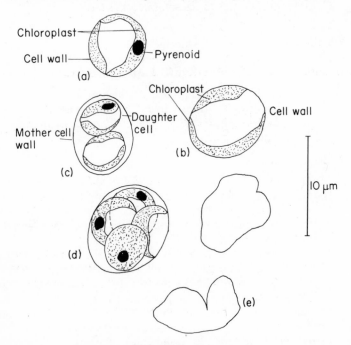

Figure 3.5 *Stages of asexual reproduction*

(a) Single cell with pyrenoid visible

(b) Single cell in different view

(c) Mother cell plus two daughter cells

(d) Mother cell plus four daughter cells

(e) Mother cell wall fragments after release of daughter cells

EXPERIMENT 3.3

THE LIFE CYCLE OF A SLIME MOUND—*DICTYOSTELIUM DISCOIDEUM*

THEORY

The slime moulds are a group of organisms that are found most commonly growing in damp situations where there is plenty of organic material. Thus decaying logs and leaf litter in woods are habitats in which slime moulds can be found.

The slime moulds are a fascinating group as they show characteristics that are traditionally associated with both the 'animal' and 'plant' kingdoms. This alone is one good reason for studying them, but in addition these organisms are widely used as experimental material in research into protoplasmic streaming, cell movement and cell aggregation, cell differentiation, etc.

In this experiment you will observe the life cycle of a cellular slime mould—*Dictyostelium discoideum*.

PROCEDURE

1. You will be provided with a slope culture of *D. discoideum*.
2. Take four sterile Petri dishes and 4×15 cm³ rabbit dung agar (RDA) (or Corn Meal Agar), from the 45°C water bath. Pour four RDA plates. Allow to set and label RDA.
3. Flame-sterilise an inoculating loop and cool. Aseptically take up some of the *D. discoideum* culture and streak it on the surface of the first RDA plate. Label Experiment 3.3, Group No. . . . , D. disc., plate 1.
4. Repeat Stage 3 and inoculate the remaining three plates. Label them plates 2, 3 and 4.
5. Incubate the plates at 25°C.

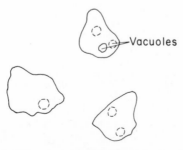

Figure 3.6 *Myxamoebae of* Dictyostelium discoideum

EXAMINATION OF THE SLIME MOULD

6. After 1 day take one of the inoculated plates and remove a small square of agar. Mount in a drop of water on a clean glass slide.

7. Examine under the high-power lens (or oil-immersion) of your microscope for minute **amoebae**—usually called **myxamoebae** (see *Figure 3.6*). These ameobae feed on bacteria on the agar surface. The bacteria were added with the original inoculum.

8. After a few days incubation at 25°C, examine the plates, and repeat this examination daily with a binocular microscope (dissection microscope or hand lens). Look for areas on the plate where the amoebae have started to migrate in streams and may have produced heaps of cells on the agar surface. These heaps are known as **pseudoplasmodia**—see *Figure 3.7*.

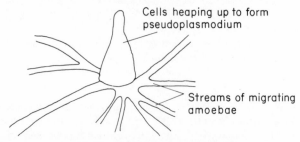

Figure 3.7 *Formation of pseudoplusmodium*

9. Continue to examine the plates—daily if possible.

Note and record the development of the pseudoplasmodium. You will see that the pseudoplasmodium topples over and becomes a cartridge-shaped structure that migrates across the plate. The slug produces a slime trail as it migrates across the plate and often leaves a few of the amoebae behind, in the slime (see *Figure 3.8*). Eventually the slug transforms into a

Figure 3.8 *Migrating pseudoplusmodium*

fruit body—**sorocarp**, which is made up of a basal disc and a stalk—the **sorophore**, at the top of which is a mass of spores embedded in slime (sometimes called a sporangium) (see *Figure 3.9*).

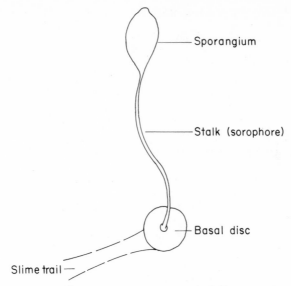

Figure 3.9 *Fruit body of* D. discoideum

10. It is difficult to mount the fruit bodies without breaking them up, but it is worth while mounting one to observe the structure of the stalk and the spores. Carefully remove a fruit body with a sterile scalpel and mount in water. Examine under your microscope.

The stalk is made up of an outer cellulose sheath and an inner cellular structure (see *Figure 3.10*). The spores (see *Figure 3.11*), also have a cellulose wall. Given the correct environmental conditions the spores will germinate to produce new myxamoebae.

QUESTIONS

1. What features of *Dictyostelium discoideum* would you class as (*a*) 'animal' -like and (*b*) 'plant' -like?
2. Produce a diagrammatic summary of the life cycle of this organism.

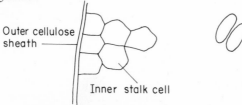

Figure 3.10 *Optical section of stalk of sorocarp*   Figure 3.11 *Spores*

3. What factors do you think may stimulate aggregation of the myxamoebae?
4. Suggest a mechanism to account for the phenomenon of aggregation.
5. How do you think the pseudoplasmodium moves?

EXPERIMENT 3.4

THE LIFE CYCLE OF A NON-MYCELIAL ASCOMYCETE-YEAST

THEORY

Yeasts are fungi belonging to the Class Ascomycetes. Some are grouped under the Order Saccharomycetaceae, i.e. yeasts which reproduce asexually by budding, for example brewers' or bakers' yeast *Saccharomyces cereviseae*. They are essentially unicellular organisms which occur naturally on the surface of fruits, although it is thought that they have arisen from mycelial forms of fungi.

PROCEDURE

You will be provided with a culture of yeast in 2% malt extract.

Stages of asexual reproduction

1. Remove a small drop of the culture of actively growing yeast cells. Mount it on a clean glass slide and examine using the high power of the microscope.

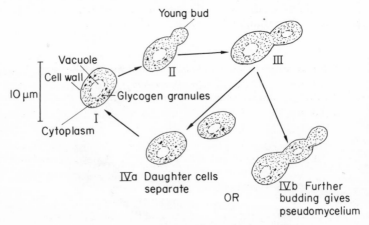

Figure 3.12 *Stages in asexual reproduction—yeast*

2. Refer to *Figure 3.12* and then look for stages of budding (asexual reproduction). Attempt to find a **pseudomycelium**, i.e. a chain of cells produced by budding. Draw all the stages you can find.

Food storage in the cells

3. Having familiarised yourself with the appearance of yeast cells, mount another drop of the yeast culture on a clean slide. Add to this a drop of iodine in potassium iodide solution and cover with a cover slip. Examine under the high power microscope.

> Record any changes occurring in the cells.
> Do the yeast cells contain food reserve?
> If so, can you suggest the form the food reserve is in?

Formation of resting spores

4. Pour a sterile plate of 0·5% sodium acetate agar. This medium is a starvation medium designed to induce the organism to produce resting spores.
5. Flame sterilise an inoculating loop and allow to cool. Thickly streak the yeast culture across the sodium acetate agar plate as shown in *Figure 3.13*.

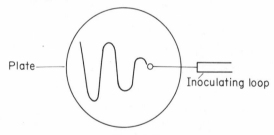

Plate

Inoculating loop

Figure 3.13

Label your plate with the experiment number, your initials and the name of the organism (*Sacc. cer.*), and leave for 1 week at room temperature.

6. After 1 week make a permanent preparation of resting spores of yeast.
   (*a*) Place a drop of distilled water on a clean grease-free slide.
   (*b*) Using a flame-sterilised loop remove some yeast from the sodium acetate agar plate prepared above.
   (*c*) Mix the yeast with the distilled water to make a suspension and spread this out to form a thin film of yeast cells.
   (*d*) Air-dry your film. MAKE SURE IT IS COMPLETELY DRY.

(e) Fix the slide by passing it 2–3 times through a cool Bunsen flame. Do not overheat or you will destroy the cells. The slide should never become too hot to touch.

CARRY OUT ALL THE SUBSEQUENT STAGES ON A STAINING RACK OVER A BENCH SINK.

(f) Cover the material with 5% aqueous malachite green solution. Gently heat the slide for 30 s as follows: make a twist of cotton wool on the end of a glass rod and soak it in 90% ethanol. Light the alcohol and wave the flame to and fro under the slide. Do not allow the liquid on the slide to boil.

(g) Rinse off with tap water.

(h) Counterstain with 0·5% aqueous safranin solution for 1 min.

(i) Wash off the stain with tap water. Air-dry the slide.

(j) Mount in Canada balsam.

7. Examine your preparation. The resting spores are called **ascospores**. Refer to *Figure 3.14* and then identify ascospores, which will be stained green. Ordinary vegetative cells stain red. Look for groups (tetrads) of spores inside the original mother cell wall, now called the **ascus**. Draw stages in ascospore formation.

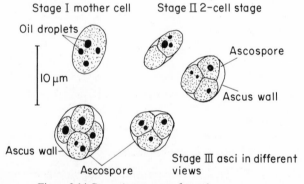

Figure 3.14 *Stages in ascospore formation—yeast*

Sexual reproduction stages

8. Take a tube containing 10 cm³ sterile 2% malt extract.

9. Obtain cultures of α- and β-strains of *Saccharomyces cerevisiae*. These are compatible haploid mating strains.

10. Flame-sterilise an inoculating loop and cool. Take up a loopful of the *a*-strain and aseptically transfer to the 2% malt extract.

11. Repeat Stage 10 using the β-strain. Mix throughly and incubate at room temperature.

12. Transfer one drop of the suspension immediately to a clean glass slide, using a sterile Pasteur pipette. Examine the cells under the high power of a microscope. Look for stages of cell fusion as shown in *Figure 3.15*. Draw the cells you can see.
13. Repeat Stage 12 after 24 h and look closely for stages of cell fusion. Draw as many stages as you can see.
14. Complete your study of the yeast life cycle by drawing a diagram showing all phases of the cycle.

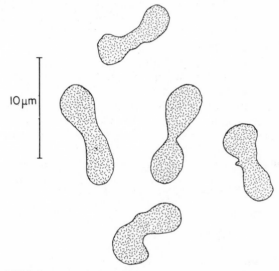

Figure 3.15 *Stages in asexual reproduction—fusion of haploid-compatible yeast cells*

EXPERIMENT 3.5

THE LIFE CYCLE OF *MUCOR HIEMALIS*

Mucor hiemalis is a mycelial Phycomycete fungus which belongs to the Order Zygomycetes. The Phycomycetes are fungi which are aseptate at least in the young state and the Zygomycetes are those members that produce a resting spore called a zygospore. This experiment involves studying the various stages of the life cycle of this organism.

PROCEDURE

A. Examining the mycelium

1. Take a culture of *Mucor hiemalis* on Malt Extract Agar (MA), in which the colony has not reached the sides of the dish.

2. Flame-sterilise a scalpel and cool. Remove a small square of agar plus mycelium from the edge of the growing colony (see *Flow Diagram 3.2 (1)* on page 74).
3. Mount this in water on a clean glass slide. Examine under the low-power objective of your microscope. Make sketches to show how the hyphae branch, (see *Figure 3.16*).

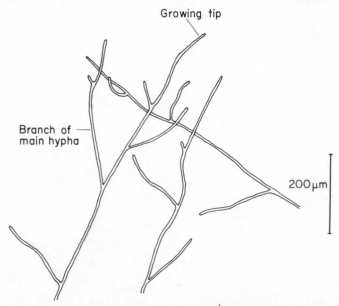

Figure 3.16 *Young mycelium of* Mucor hiemalis

4. Examine, under the high-power objective, some young hyphae. Draw a young hypha or hyphae to show internal details. Is this organism of recognisable units? (Refer to *Figure 3.17*.)
5. Irrigate your preparation with iodine in potassium iodide using the blotting paper technique described in Experiment No. 3.2 on page 63.
6. Refer to *Figure 3.17* and explain what happened when you added the iodine in potassium iodide. Make drawings to show any changes.

B. The older hyphae and chlamydospores

6. Using a flame-sterilised, cooled scalpel, cut out a very small triangle of agar + mycelium from the centre of the colony used in procedure Stage 1 (see *Flow Diagram 3.2(1)*).

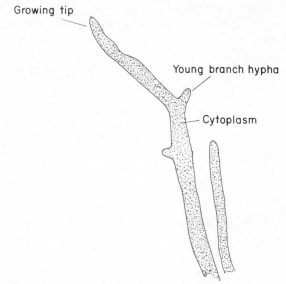

Figure 3.17 *Young hyphae and contents*

7. Squash this triangle on a slide together with two drops of Sudan blue. Examine at low and high powers.
8. Sudan blue is taken up by fat droplets. Refer to *Figure 3.18*. Are there fat droplets inside the hyphae? Draw a hypha showing fat droplets.

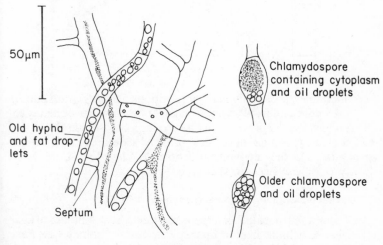

Figure 3.18 *Older mycelium*

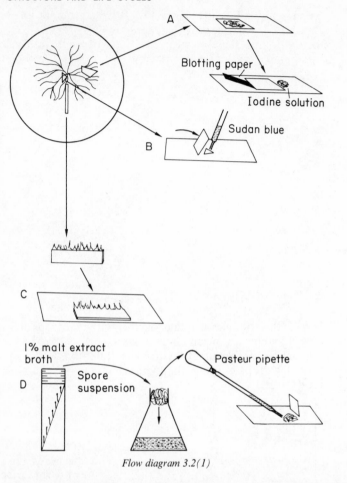

A

Blotting paper

Iodine solution

Sudan blue

B

C

1% malt extract
broth

D

Spore
suspension

Pasteur pipette

Flow diagram 3.2(1)

9. Examine the older hyphae for partitions, i.e. cross walls cutting the hyphae off from (*a*) empty lateral branches, or (*b*) chlamydospores. (See *Figure 3.18*.) Draw examples of (*a*) and (*b*).

The chlamydospores have thick walls and are more or less rounded structures. Do they contain fat droplets? What do you think the function of these structures might be?

C. Asexual reproduction—sporangium development

10. Using a sharp scalpel or a razor blade cut a very thin slice of agar plus mycelium, from a diameter across the colony (see *Flow Diagram 3.2(I)*). Mount the slice in water on its side.

Examine it under the low power of the microscope and look for erect hyphae. Start at the youngest end on your agar slice and try to find stages in the formation of sporangia. Refer to *Figure 3.19*, and look for hyphae with a terminal swollen tip;

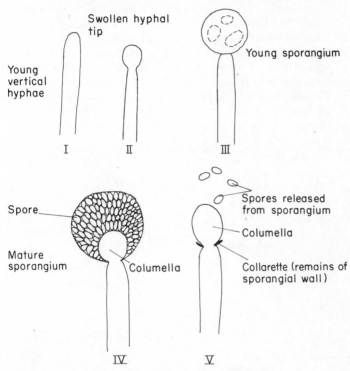

Figure 3.19 *Stages in sporangium formation*

hyphae with the sporangium apparently intact but with the protoplasm not yet organised into spores; and for sporangia plus spores. Draw as many stages as you can find.

11. Older sporangia may have released their spores. Why can you not find wall fragments? Draw any remaining structures.

12. When sporangia are mature in *M. hiemalis*, they are found in a mucilaginous drop on the top of the sporangiophore (sporangium bearing hypha). On exposure to dry air, however, water from the drop evaporates leaving the spores cemented to the columella and to one another. How do you think these spores can be dispersed under natural conditions? What is the function of these spores?

D. Asexual reproduction—spore germination

13. Take a slope culture of *M. hiemalis* which has sporangia.
14. Add to the culture about 10 cm³ of sterile 1 % MA Broth.
15. Shake gently to produce a spore suspension.
16. Transfer this suspension to a small (50 or 100 cm³) conical flask which has been plugged with cotton wool and sterilised (see Chapter 1 for making cotton wool plugs).
17. Using a Pasteur pipette remove a small drop of suspension and mount it in 1 drop of 0·1 % methylene blue. Examine under the high-power or oil-immersion lens of your microscope.
18. Repeat procedure Stage 17, after 2 h, 4 h, and 8 h. Draw stages of spore germination at each time (see *Figure 3.20*). What changes appear to occur as the spore germinates?

Spores swell and become vacuolated

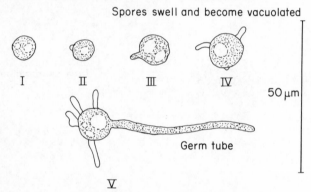

50 μm

Germture

Figure 3.20 *Stages in spore germination*

E. Sexual reproduction

19. Obtain two cultures of *M. hiemalis*. One which is a + strain and the other a − strain.
20. Pour three sterile MA plates and label them Experiment 3.5.E and with your group number.
21. Flame-sterilise a scalpel and cool. Remove a small piece of agar together with some + strain mycelium from the *M. hiemalis* + culture. Place this piece face down on the surface of the first plate so that the hyphae contact the agar surface.
22. Repeat Stage 21 with the − strain culture. Place the two pieces approximately 3 cm apart. Label the plate + and − on the bottom of the dish over the inocula.
23. Repeat Stages 21 and 22 but prepare plates containing + and + inocula; and − and − inocula. Carefully label all your plates.
24. Leave the plates at room temperature for about 5 days.

25. Flame-sterilise a scalpel and remove some mycelium and agar from the centre of the first plate (see *Flow Diagram 3.2(2)*). Mount this in water and examine for stages of sexual reproduction. Refer to *Figure 3.21* and familiarise yourself with stages you are likely to find. Draw as many stages as you can find under the microscope.

26. Repeat your examination on the other two plates prepared in Stage 23. Compare your findings with those from the +/− plate. Can you explain any differences?

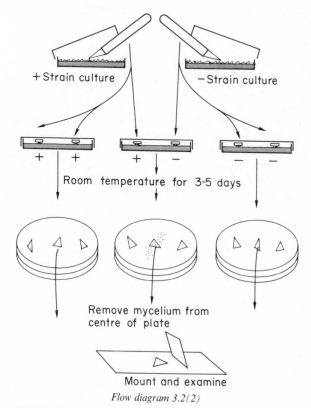

+ Strain culture   − Strain culture

+   +   +   −   −   −

Room temperature for 3-5 days

Remove mycelium from centre of plate

Mount and examine

Flow diagram 3.2(2)

F. Zygospore germination

Zygospores are difficult to germinate in the laboratory, but when they do germinate asexual sporangia are produced.

The nuclei of the gametangia involved in conjugation are haploid. The asexual spores produced after the zygospores have germinated are also haploid.

What nuclear changes do you think occur from the time of the formation of the zygospore to the time of asexual spores being produced?

Suggest a function for the zygospore in the life cycle of *Mucor* sp. Draw a diagram of the life cycle of *M. hiemalis*.

EXPERIMENT 3.6

AN EXAMINATION OF TWO MYCELIAL ASCOMYCETES— *ASPERGILLUS (EUROTIUM)* AND *PENICILLIUM*

THEORY

The class Ascomycetes is the largest of the classes of fungi. Its members are characterised by the production of a spore containing a sac-like structure called an *ascus*. This is produced after sexual reproduction. Early in its development the ascus contains two haploid nuclei, which fuse giving a diploid nucleus. This immediately undergoes meiosis forming four haploid nuclei. In most members of the class (except, for example, the Order Saccharomycetaceae— budding yeasts, see Experiment 3.4), these four haploid nuclei undergo mitosis to give eight haploid nuclei. Cytoplasm is then organised around the nuclei to form ascospores. These events are summarised in *Figure 5.1*, page 104, for *Sordaria fimicola*.

In a few members of this class either the ascus stage is rarely seen or is apparently completely absent. In these organisms the asexual phase dominates the life cycle. Two such organisms are *Aspergillus* species (*Eurotium*), and *Penicillium* species, which you will investigate in the following experiment.

PROCEDURE

You will be provided with slope cultures of *Penicillium* and *Aspergillus* sp. both bearing spores.

Part A

1. Take two sterile Petri dishes, an inoculating loop and Bunsen, also the above cultures.
2. Take two 15 cm³ amounts of sterile quarter-strength malt extract agar from the 45°C water-bath
3. Pour two quarter-strength MA plates and allow to cool and set.
4. Flame-sterilise the loop and cool. Aseptically take up some of the spores of *Penicillium* from the slope culture. Inoculate the first plate by streaking the spores across the surface. Label your plate Experiment 3.6, Group No. ....., and *Penicillium*.

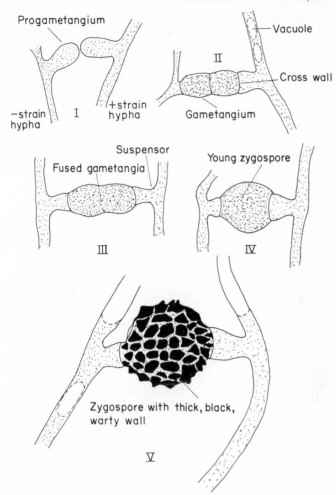

Figure 3.21 *Stages in sexual reproduction of* Mucor hiemalis

5. Repeat Stages 3 and 4 for *Aspergillus* and label accordingly. Label the other plates as controls.
6. Incubate at room temperature for 1 week.

Part B

7. Now pour a plate of sterile MA containing additional carbohydrate (e.g. 20 g sucrose per 100 cm^3).
8. Inoculate with spores of *Aspergillus* as in Part A. Incubate at 25–30°C for 10 days.

Part C. Mycelial structure and asexual reproduction

9. In *Penicillium*
   Take the plate of *Penicillium* prepared in Part A and very carefully, using a flame-sterilised and cooled scalpel, cut out a 1 cm² area from the edge of the developing *Penicillium* colony (see *Flow Diagram 3.3*).

Penicillium culture

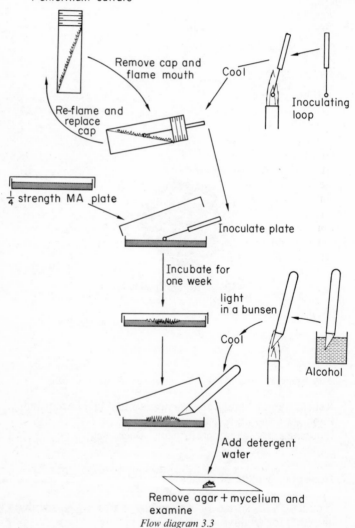

Flow diagram 3.3

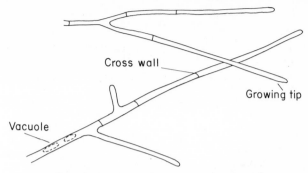

Figure 3.22 *Young hyphae of* Penicillium notatum

10. Mount the square in some water to which a few drops of liquid detergent have been added (e.g. put 4–5 drops of detergent into 20 cm³ of water).
11. Using the high-power objective of the microscope focus carefully on the hyphae on the surface of the agar. Draw a few hyphae from the growing edge of the colony to show branching and septa (cross-walls). Refer to *Figure 3.22*.

One interesting feature of the mycelium in Ascomycetes is the hyphal fusion which occurs between neighbouring hyphae. This is called **anastomosis** and produces a 3D hypal network.

12. Try to find a few stages of hyphal anastomosis and draw them. Refer to *Figure 3.23*.

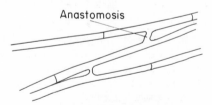

Figure 3.23 *Hyphal anastomosis*

13. Look at *Figure 3.24* and then find a conidiophore bearing chains of conidia. Draw the structure you find. Notice that the spores are not produced inside sporangia (cf. Mucor—Experiment 3.5), but are cut off in succession from the ends of the phialides. What do you think the method of spore dispersal is in this organism?

14. In *Aspergillus*
    Take the plate of *Aspergillus* prepared in Part A and repeat Stages 9–12 using *Aspergillus*.

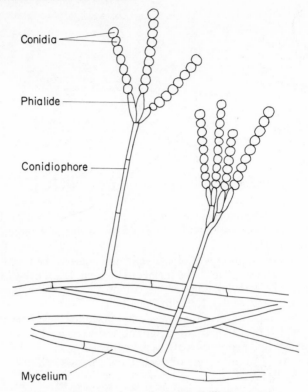

Figure 3.24 *A sexual reproduction*–Penicillium notatum

15. Find a conidiophore with chains of conidia present. Refer to *Figure 3.25*, and then draw your structure. Note the basic similarities of structure with the conidia of *Penicillium*.
    Note: Do not confuse the swollen head of the conidiophore in *Aspergillus* with a sporangium.

Part D. Sexual reproduction

Take the plates of *Penicillium* and *Aspergillus* prepared in Part B. The extra carbohydrate, together with the higher incubation temperature induces the formation of sexual stages.

In *Aspergillus* the sexual stage is called *Eurotium*. The asci are produced in a structure called a **cleistothecium**, which has no opening via which spores are released (compare this with the perithecium of *Sordaria*—Experiment 5.1). How do you think the ascospores are released from the cleistothecium?

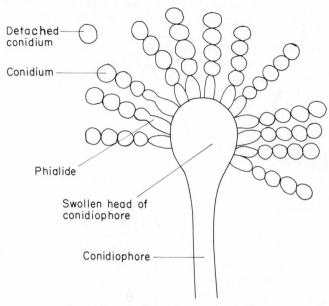

Figure 3.25 *Asexual reproduction*—Aspergillus *sp.*

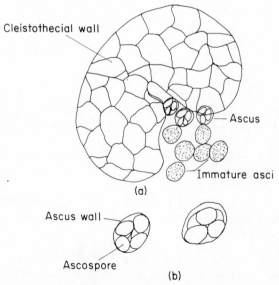

Figure 3.26 (a) *Crushed cleistothecium*
(b) *Asci*

16. Take a small amount of the mycelium plus cleistothecia from the plate of *Aspergillus*, and mount it in water on a clean glass slide.
17. Crush the material by squashing it, and try to find cleistothecia containing asci. You should find asci plus ascospores separate from their cleistothecia (refer to *Figure 3.26*). Make a drawing from your material.

EXPERIMENT 3.7

THE LIFE CYCLE OF A BASIDIOMYCETE—*AGARICUS BISPORUS*

THEORY

The Basidiomycetes are probably the most widely recognised group of micro-organisms because of the relatively large size of their fruit bodies. The group includes the very familiar toadstools, bracket fungi, puff balls, and stinkhorns, and also the economically important edible mushroom— *Agaricus bisporus*.

Not all of the Basidiomycetes are conspicuous however. Two groups, the smuts and the rusts, that do not produce fruit bodies, are parasites of higher plants.

The fruit body of the group is essentially a structure for the production and discharge of sexually produced **basidiospores**. These spores are produced on special club-shaped cells called **basidia**, and together they constitute the major characteristic feature of the group.

The fruit body is, however, only part of the organism, and in many members of the group it is a minor part of the organism, only lasting for a few days during the autumn. The major part of these organisms is the mycelium, which although less conspicuous, consists of well developed hyphae that form a network on or in the substrate on which the organism is growing, and absorb nutrients from this substrate.

In many ways the structures found in the cultivated mushroom are not characteristic of the Basidiomycetes as a whole, and therefore in this experiment you will consider the mycelium of another organism—*Polystictus*, at the same time. *Polystictus* is a very common bracket fungus found growing on dead logs.

PROCEDURE

You will be provided with a slope culture of *Polystictus*.
1. Take the slope culture and a scalpel. Obtain a sterile Petri dish and 15 cm³ of sterile potato dextrose agar (PDA) from the 45°C water-bath.

2. Pour a PDA plate and allow to cool and set. Label PDA.
3. Using the aseptic technique, remove a small piece of the *Polystictus* mycelium from the culture and place it face down on the centre of your PDA plate. Label the dish with Experiment No. 3.7, Group No., and *Polystictus*.
4. Incubate the plate at 25°C for 5 days.

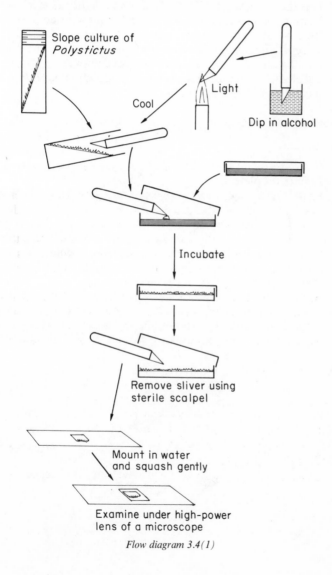

Flow diagram 3.4(1)

You will be provided with a young cultivated mushroom.

5. Check that the mushroom fruit body has not yet opened to expose the gills.

6. Take a sterile Petri dish, a pair of forceps and 15 cm³ of sterile malt extract agar, the latter from the 45°C water bath. Pour a MA plate and allow to cool and set.

7. Dip the forceps into 100% alcohol and light the alcohol in a Bunsen (as in sterilising of a scalpel). Allow to cool for 30 s. Remove the gill covering with the forceps. Re-sterilise the forceps, and again cool. Remove some of the gill and place it into the centre of the MA plate prepared above.

8. Label your plate MA, Experiment 3.7, Group No., *A. bisporus*. Incubate at 25°C for 10 days.

EXAMINING THE MYCELIUM

A. *Polystictus*

1. Cut out a small sliver (thin slice) of agar + mycelium from the *Polystictus* plate.

2. Place it on its side in a drop of water on a glass slide. Cover with a cover slip and gently squash.

3. Examine under the high-power or oil-immersion lens of your microscope. Notice the branched hyphae with well defined cross walls (in this way the mycelium resembles an Ascomycete). Look for clamp connections—refer to *Figure 3.27*, and try to find stages in the formation of clamp connections.

Note: Clamp connections are a distinct feature of many Basidiomycetes. A mycelium with clamp connections has hyphal sections separated by cross-walls, and each section contains a pair of nuclei. This is the **dikaryotic** condition. This situation is

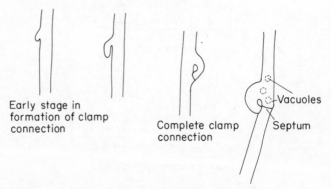

Early stage in
formation of clamp
connection

Complete clamp
connection

Vacuoles

Septum

Figure 3.27 *Clamp connections in* Polystictus

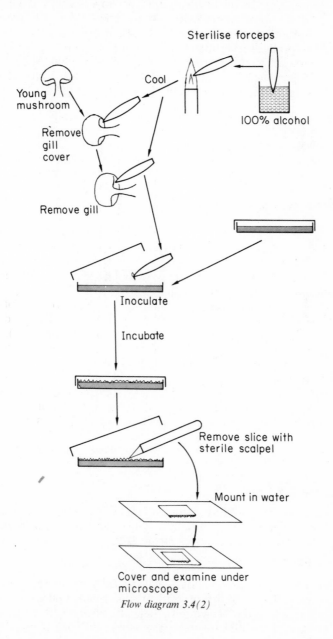

Sterilise forceps

100% alcohol

Young mushroom

Cool

Remove gill cover

Remove gill

Inoculate

Incubate

Remove slice with sterile scalpel

Mount in water

Cover and examine under microscope

Flow diagram 3.4(2)

produced by the fusion of branched mycelia each of which was made up of uninucleate sections (cells), i.e. was in the **monokaryotic** condition. The dikaryotic mycelia are the only ones to produce fruit bodies, and the purpose of clamp connections appears to be to ensure that each of the hyphal sections in the dikaryotic mycelium contains two nuclei, one from each of the original monokaryotic mycelia (see *Figure 3.28*).

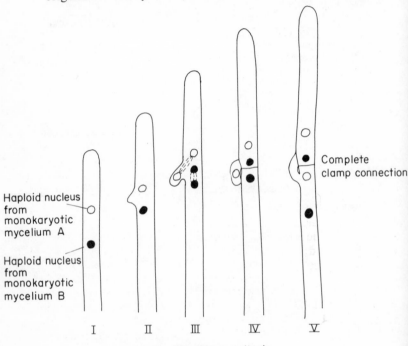

Figure 3.28 *Stages in formation of clamp connections*

B. *Agaricus bisporus*

Use the same procedure as for *Polystictus*. Note that *Agaricus* does not have clamp connections, thus its mycelium shows only branching hyphae with cross-walls—see *Figure 3.29*.

EXAMINING THE FRUIT BODY OF *AGARICUS BISPORUS*

1. Take a young fruit body. Cut it vertically into two and examine. Note that it consists of: stalk or **stipe**, a cap or **pileus**, and under the pileus the brown gills. Refer to *Figure 3.30*. Look at the base of the stipe for the connection to the feeding mycelium.

Branched septate hyphae

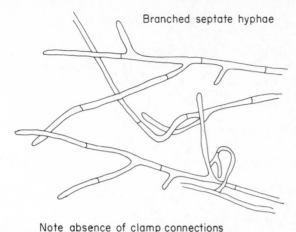

Note absence of clamp connections

Figure 3.29 Agaricus bisporus *mycelium*

Usually near the top of the stipe you can see a ring of tissue, the **annulus**, that is a remnant of the membrane (velum) which in the young fruit body encloses the gills in a cavity by joining the edge of the pileus to the stipe (see *Figure 3.31*). Make a fully labelled diagram of the fruit body.

2. Take a very sharp botanical razor or preferably a new razor blade in a holder and cut across the pileus at right angles to your first cut so that the gills can now be seen in cross-section. Cut a very thin section of the gills so that they can be examined under the microscope. DO NOT try to cut a complete section—a small, thin fragment is sufficient. Remove the section using a camel-hair brush taking care not to damage the section. DO NOT use forceps or needles. Add a drop of water and cover with a cover slip. Examine under the high-power lens of your microscope.

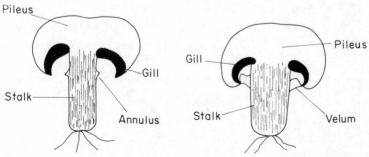

Figure 3.30 *L.S. mature fruit body*   Figure 3.31 *L.S. young fruit body*

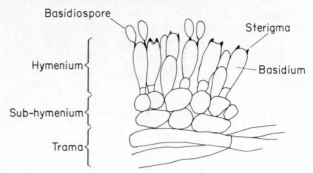

Figure 3.32 *High-power detail of T.S. gill*

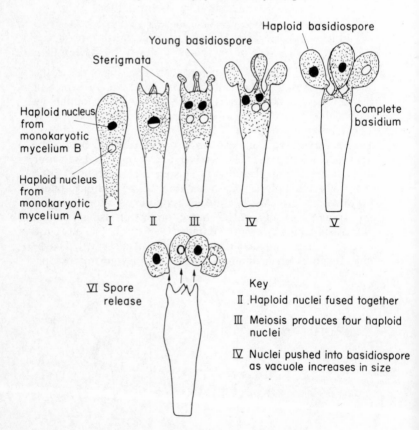

Figure 3.33 *Stages in basidium formation*

Note. That the gills are made of an inner section of parallel hyphae —the **trama**, and on the outside club-shaped cells at right angles to the trama—the **basidia**. See *Figure 3.32*. The basidia have projections called **sterigmata** at the ends of which are found the haploid basidiospores. Most Basidiomycetes have four spores per basidium but notice that *Agaricus bisporus* (as its name suggests) has only two basidiospores per basidium.

Look at *Figure 3.33*, and then try to find as many stages of basidial formation as you can on your section. Draw all the stages you can find.

QUESTIONS

1. How do you think the spores are released from the basidium?
2. What is the role of the fruit body in spore dispersal?
3. From your investigations above and the knowledge gained draw a life cycle diagram for a basidiomycete of the mushroom type.
4. What is the significance of the arrangements of the gills in the mushroom?

4
Growth

INTRODUCTION

The terms 'growth' and 'development', are very difficult to define, as they are by nature very complicated processes affected by a large number of interrelated variables.

If we are investigating unicellular organisms, their cell division will cause an increase in the numbers of the population, but for a single multicellular organism, cell division causes increase in the individual's size, and may involve morphological and physiological changes.

At best we can probably apply the term **growth** to a quantitative increase in the size of either a population of unicells or of a multicellular organism. The qualitative changes can be referred to as **development**.

We mentioned above the effect of variables on growth and development. These include such things as: the level of available nutrients, light, temperature, etc. Many of these can be controlled in the laboratory and the resulting growth patterns from such experiments can be analysed and a mathematical description of growth under controlled conditions can be derived.

The experiments in this chapter give you the opportunity to investigate growth patterns (*a*) in a unicellular autotroph, and (*b*) in a filamentous heterotroph, and to apply some of the processes mentioned above.

EXPERIMENT 4.1

THE GROWTH OF A POPULATION OF A UNICELLULAR ORGANISM—*CHLORELLA*

THEORY

The life cycle and nutrition of the green alga, *Chlorella* sp., are investigated in Experiment Nos. 3.2 and 2.1 respectively. If you have not already carried out these experiments, reference to the information and diagrams contained in these sections will inform you about the organism *Chlorella*.

In this experiment you will examine the way in which a population of unicellular organisms grow. The organism used is *Chlorella* sp. which is grown in liquid culture in the light.

PROCEDURE

1. Look at *Figure 4.1* showing the apparatus used in this experiment. Set up a 250 cm³ conical flask with a rubber bung and glass tube leading to the pump. Pack a U-tube with non-absorbent cotton wool and put it in the two glass tubes and bungs.

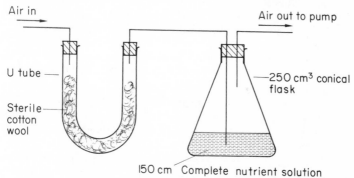

Figure 4.1 *To show apparatus for Experiment 4.1*

2. Put 150 cm³ of a complete nutrient solution (as used in Experiment 2.1, see page 42) into the 250 cm³ conical flask.
3. Autoclave the two pieces of apparatus at 10 lb/in² for 15 min. Allow them to cool inside the autoclave.
4. Remove the apparatus from the autoclave and quickly join the two parts together. Attach to a pump.
5. Take a sterile Pasteur pipette and add sufficient *Chlorella* suspension to produce a slight green tinge to the medium in the flask. The suspension you use will have been produced by culturing *Chlorella* at room temperature in the complete medium for 2–3 days in the light.
6. Switch on the pump and ensure that a good flow of air occurs. Leave the apparatus in the light.

The U-tube with its sterile cotton wool filters dust particles and contaminating organisms from the air entering the apparatus. This 'clean' air bubbles through the flask and the agitation produced by the bubbling maintains a constant circulation of nutrients and *Chlorella* cells. The medium supplies all the necessary inorganic substances for growth whilst the culture receives a constant supply of oxygen and carbon dioxide from the air.

ASSESSING THE GROWTH

7. As soon as the apparatus has been set up and is aerated, remove one drop of the medium using a sterile Pasteur pipette (*N.B.* see below, Stage 8). Place this drop on a clean glass slide and cover with a cover slip. Examine under the high power of your microscope. Count the number of *Chlorella* cells in the field of view. Looking from the side of the microscope, move the slide slightly. Examine this new field of view and count the *Chlorella* cells. Repeat this for eight more high-power fields taken at *random*. *N.B.* If the numbers of cells/10 high-power fields is less than five, increase the concentration of *Chlorella* (see Procedure, Stage 5).
8. The time at which you take your first reading is day 0. Repeat daily using the SAME PASTEUR PIPETTE to take your samples. This is essential to produce the same drop size. Also use the same size of cover slip. This together with the drop size will produce a field of comparable depth throughout the whole of the experiment. Take the time each reading is made.
9. Tabulate your results.
10. If a haemocytometer is available , you can use this to express your results as actual numbers of Chlorella cells per cm^3 of culture. (See Appendix opposite.)
11. Draw a graph of your results with time on the x axis and numbers of cells on the y axis.

QUESTIONS

1. Does the main part of your graph correspond to graph (*a*) in *Figure 4.2* below or to graph (*b*)?
   Graph (*a*) is the graph of an arithmetic series of the type 0, 1, 2, 3, 4, 5, . . . Graph (*b*) is the graph of a geometric series (or exponential series) of the type 1, 2, 4, 16, 32, . . . , or to write it another way 2^0, 2^1, 2^2, 2^3, 2^4, 2^5, . . .

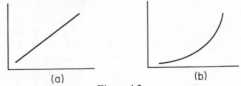

(a)   (b)

Figure 4.2

If your graph corresponds to (*b*), how can you modify your numbers to produce a straight line graph, i.e. convert the geometric series to an arithmetic one?

Plot a new graph using the method you decide upon.

2. Can you divide your second graph into a number of different phases of growth? List them and explain why each phase occurs.
3. What do you think is the outstanding phase of growth in a population of micro-organisms?
4. How does the type of growth compare with the growth of a multicellular organism, and the growth of a filamentous fungus (see Experiment 4.2).
5. How could you increase the growth rate of this organism?

Appendix to Experiment 4.1

THE USE OF A HAEMOCYTOMETER

The following data applies to haemocytometers of the improved Neubauer type.

The haemocytometer consists of a thick glass slide with four channels extending across it (see *Figure 4.3*). The central platform is made such that when the cover glass is in position the chamber has a depth of 0·1 mm.

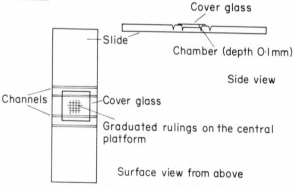

Figure 4.3 *To show a haemocytometer*

The central area of the central platform has a number of graduated rulings. These are shown in *Figure 4.4*. The area marked out is 9 mm². You will use only the central square—the one inside the ring on *Figure 4.4*. This and the other large squares are 1·0 mm² in area. This central 1 mm² area is subdivided into 25 smaller squares each having an area of 0·04 mm². The lines marking the edges of each 0·04 mm² area are triple lines (see *Figure 4.5*).

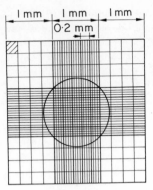

Figure 4.4

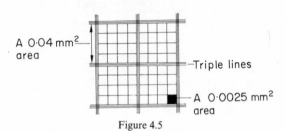

Figure 4.5

Each 0·04 mm² area (in the central area you are using) is subdivided into 16 smaller squares each with an area of 0·0025 mm².

PROCEDURE FOR USE

1. Very carefully clean the haemocytometer slide and cover slip. Slide the cover slip into position by holding the slide and cover slip as in *Figure 4.6*. Push down hard on the cover slip with both thumbs and slide it across the slide.

   Where the slide and cover slip are in close contact, a rainbow effect should be seen when the cover slip is viewed from a shallow angle. (These are Newton's rings.)

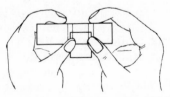

Figure 4.6

2. Now take a sterile Pasteur pipette. Thoroughly shake the culture to give an even dispersion of cells. Take up some of the culture into the pipette.

3. Place the pipette as shown in *Figure 4.7*, the end of the pipette next to the cover glass.

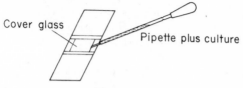

Cover glass

Pipette plus culture

Figure 4.7

Pipette out a drop of culture on the slide and by capillarity it will fill the haemocytometer chamber.

N.B. IF AIR BUBBLES BECOME TRAPPED UNDER THE GLASS or if the CHANNELS BECOME FULL OF LIQUID, CLEAN THE SLIDE AND START AGAIN.

4. Place the slide on your microscope and focus the high-power objective lens on the central area of the counting chamber (ringed *Figure 4.4*).

5. Count all the cells in each of five randomly selected 0·04 mm² areas. This is considered to be a statistically viable sample of the central area. Cells touching the triple lines demarking 0·04 mm² areas should be disregarded, but cells touching the lines dividing 0·0025 mm² areas should be counted.

If the number of cells per 0·0025 mm² area is more than approximately 5, then the cell suspension is too concentrated. Dilute the sample (by $\frac{1}{2}$ or a $\frac{1}{4}$ as needed) before counting.

6. Calculate the number of cells/cm³. Now make a high-power assessment from the same material and use the haemocytometer count to calibrate this and your other high-power assessments. Thus make only ONE haemocytometer count towards the end of the experimental period.

EXPERIMENT 4.2

GROWTH OF A FILAMENTOUS ORGANISM—*MUCOR HIEMALIS*

THEORY

Mucor hiemalis, the life cycle of which is examined in Experiment 3.5 is a Phycomycete fungus. It occurs commonly on dung, where it produces a mycelium of coarse, loosely growing hyphae. As a species of Mucor, it has been used extensively in laboratory work.

You will investigate the rate of growth of this filamentous organism across the surface of a nutrient medium.

PROCEDURE

1. Pour a sterile plate of malt extract agar (MA).
2. Take a slope of *Mucor hiemalis* in a MacCartney bottle. Add a small amount of sterile distilled water. Shake gently to produce a spore suspension.
3. Obtain a sterile Pasteur pipette and take up some of the spore suspension.
4. Place one *small* drop of the spore suspension in the centre of your sterile MA plate.
5. Leave the plate with its lid on on your bench for 10 min to let the spore drop dry.
6. Label the dish with the Experiment No. and your Group No.
7. Turn the plate over and mark the limit of the spore drop as follows: Put a glass slide on the upturned Petri dish and using this like a ruler scratch a line on the base of the dish using a mounted needle (see *Figure 4.8*).

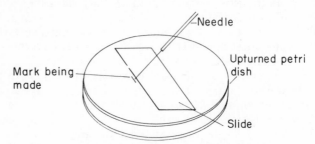

Figure 4.8 *To show marking the extent of the spore drop*

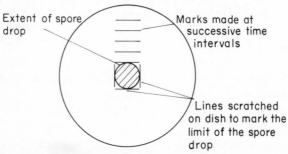

Figure 4.9 *To show marking on base of Petri dish (only one radius completed)*

Scratch a straight line to mark the outer limit of the drop. Mark the edge of the drop on all four sides (see *Figure 4.9*).

8. Repeat Stage 7 at 1 day intervals, making fresh marks to show the point to which the hyphae have grown (see *Figure 4.9*).

9. Continue Stage 8 for 3–4 days.

10. Measure the distances from the edge of the spore inoculum to each time interval mark. Use, if possible, a travelling microscope. (If not, use an accurate clear plastic ruler.)

11. Tabulate time against the distances the hyphae have grown across the plate.

12. Calculate a mean result for each time interval. If available, collect the mean results from other groups and calculate mean values for the class results.

13. Plot a graph of your results using distances (i.e. dependent variable) on the y axis and time (i.e. independent variable) on the x axis.

QUESTIONS

1. Why is growth initially very slow?
2. What is the shape of the main part of the graph.
3. What does the graph tell us about the growth of *Mucor hiemalis*?
4. Do you expect this type of growth to be maintained in this experimental situation?
5. Explain the growth pattern based on what you know about the structure of this organism. Refer to the life cycle diagram you produced in Experiment 3.5.
6. What factors do you think will influence the rate of growth on an agar plate?
7. Is increase in diameter a good measure of growth?
8. Derive a mathematical expression to describe the growth of this organism under experimental conditions.
9. What is the growth rate of the organism in micrometres per hour?

5
Genetics

INTRODUCTION

In the last 10–20 years our knowledge of genetics has increased enormously, mainly as a result of research which has been carried out using micro-organisms.

Genetics is now a subject which is being investigated at the molecular level, and micro-organisms, particularly bacteria and bacteriophage viruses are being used to elucidate, amongst other things, the precise functioning of the 'gene'.

It is therefore appropriate that we should consider some aspects of genetics using micro-organisms. Thus in this chapter there are three experiments, one of which gives clear evidence for the ideas of Mendel and also involves the more recent concepts of gene segregation, recombination and crossing over. The other two experiments are concerned with (a) gene mutation, and (b) the possible functioning of a gene.

All three experiments are very relevant to present-day knowledge, for although the ideas that Mendel put forward about heredity are well accepted, it has become apparent more recently that the genetic material—DNA (deoxyribonucleic acid) is not totally stable. Questions such as, 'How stable is DNA?', and 'What sort of changes occur?', 'How often do these changes occur?', and 'What induces them?', and 'How does DNA control the functioning of the cell?'— are questions you should be thinking about as you carry out the experiments in this chapter.

EXPERIMENT 5.1

TO DEMONSTRATE GENE SEGREGATION AND CROSSING OVER USING AN ASCOMYCETE

THEORY

The 'flask fungi', or Pyrenomycetes are characterised by the formation of asci in flask-like structures called **perithecia**.

Two fungi of this group are *Neurospora* sp. and *Sordaria* sp. Both of these organisms have been used extensively in genetic research into the behaviour of genes during meiosis (reduction division), and gene control of cell activities.

In this experiment you will use *Sordaria fimicola*. Two strains are involved, one which produces black ascospores and another which produces white (colourless) ascospores. The two strains are mated, i.e. allowed to sexually reproduce with one another, and the resulting asci are examined. The appearance of the offspring spores produced as a result of sexual reproduction will depend upon the genotype of each spore.

PROCEDURE

Two strains of *Sordaria fimicola*, $C_7 H^+$ producing black ascospores and $C_7 H^-$ producing white ascospores, are provided on plates of potato dextrose agar (PDA) or on malt extract agar (MA).

1. Take a sterile Petri dish and $15 \, cm^3$ of sterile maize, yeast extract agar (MYA) from the 45°C water-bath.
2. Pour a plate of MYA. Allow to cool, invert and label MYA.
3. Dip a scalpel into industrial alcohol and light the alcohol, holding the scalpel blade-end down. When it has finished burning, allow 30–40 s for the blade to cool.
4. Use a flame-sterilised scalpel to cut out a small triangular portion of agar plus fungus from the $C_7 H^+$ culture provided. Quickly transfer it to your plate of MYA, and place the triangle face down, i.e. fungal mycelium next to the MYA, on one side of the plate (see *Flow Diagram 5.1*).
5. Repeat Stage 3 to re-sterilise the scalpel and then repeat Stage 4 BUT using the $C_7 H^-$ culture and placing the inoculum on the OPPOSITE side of the plate about 5 cm from the first inoculum (see *Flow Diagram 5.1*).
6. Invert the plate and mark the inocula + and − as appropriate. Also label the dish with the Experiment No. and your Group No.
7. Leave the plates at room temperature for 1 week.

Assessment

8. Look on your plates for black perithecia, in the mid-line between the two inoculation points (see *Flow Diagram 5.1*)
   N.B. All perithecia regardless of the colours of the ascospores, are black on the outside.
9. Using fine forceps, gently remove one or two perithecia from this mid-line area.
10. Mount on a clean glass slide in a drop of 50% ethylene glycol, and examine under the microscope.

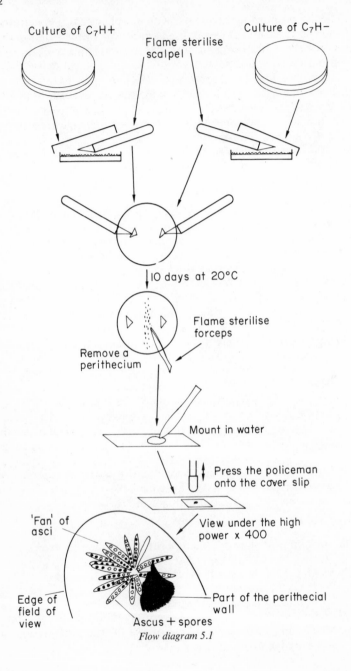

Flow diagram 5.1

11. Gently press on the cover slip with a 'policeman' (i.e. with the rubber cover on the end of a glass rod) or if necessary tap the cover slip with the 'policeman'. You should produce by this process a burst perithecium with a fan of asci outside it (see *Flow Diagram 5.1*).

12. There should be six different types of arrangement of ascospores, with the spores in pairs or fours. Make sketches of as many different asci as you can find, and make the distribution of spores clear.

N.B. Grouping of spores in other than two's or four's is not normal.

13. Draw a table of your results as in *Table 5.1*.

Table 5.1 RESULTS TABLE

Ascus type	Numbers occurring per perithecium a b c d e f g h . . .								Total
1. 4 black/4 white									
2.									
3.									
4.									
5.									
6.									

14. Collect all the results from the class and calculate total counts for each ascus type.

QUESTIONS

1. All the ascospores in the asci which resulted from the mating of these two strain are either black or white. None, for example, are grey. What does this suggest with regard to the nature of genes? What happens to the genes when the haploid spores are formed?

2. By reference to the life cycle diagram in *Figure 5.1*, describe the main differences between the life cycle of *Sordaria*, and the life cycle of higher plants and animals.

3.(*a*) Represent the chromosomes involved in this ascospore colour difference as straight lines and the genes as letters, thus:

$$B \dashv \vdash b$$

Let B be the gene for black spores and b the gene for white spores. Using these symbols show how the six ascus combinations are produced.

104

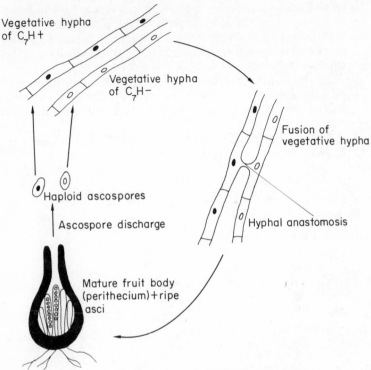

Figure 5.1 (a) *Life cycle of* Sordaria fimicola

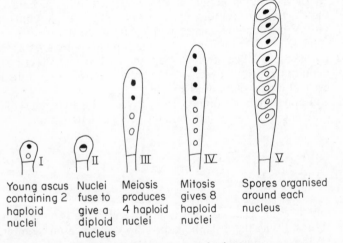

I	II	III	IV	V
Young ascus containing 2 haploid nuclei	Nuclei fuse to give a diploid nucleus	Meiosis produces 4 haploid nuclei	Mitosis gives 8 haploid nuclei	Spores organised around each nucleus

Figure 5.1 (b) *Stages in ascus development*

3. (b) Which spore combinations are produced by crossing over and thus segregation of the genes in the second meiotic division?

4. In Stage 14 above you calculated how many of each ascus type had been observed. Use this data to calculate the percentage produced by crossing over. What is the cross-over value?

EXPERIMENT 5.2

SELECTING BACTERIAL MUTANTS BY USING ANTIBIOTICS

THEORY

Two non-pathogenic bacteria, *Serratia marcescens*, a red bacterium, and *Sarcina lutea*, a yellow bacterium, are found in soil, water and the air. A third *Escherichia coli* occurs in the intestine of man and other mammals. This latter bacterium has some pathogenic strains.

Antibiotics are substances produced by living organisms which inhibit the growth of other organisms. Streptomycin is an antibiotic which is produced by the organism *Streptomyces griseus*. It is widely used as an anti-bacterial agent in medicine and other fields. The bacteria used in this experiment are usually sensitive to streptomycin. related nutritional mutants are being tested.

The experiments you will carry out investigates any genetic changes (mutations) which may occur in the bacteria, with regard to their streptomycin sensitivity.

PROCEDURE

Part 1

You will be provided with a range of nutrient agars containing increasing amounts of streptomycin from 01–00 µg/cm^3. These will be in stages as follows 0.1, 1.0, 5.0, 10.0, and 100 µg/cm^3.

1. Pour five plates using sterile Petri dishes and the sterilised agars, so that you have a representative dish for each of the streptomycin concentrations.

2. Label each of your plates carefully as follows:
   Experiment No. 5.2: the name of the organism used, the streptomycin concentration.

3. Flame-sterilise an inoculating loop and allow to cool.

4. You are provided with a culture of *Escherichia coli* in a nutrient broth suspension. Take up a loopful of this suspension and streak it over the first plate.

5. Repeat Stages 3 and 4 for all five plates.

6. Incubate the plates at 37°C for 48 h.

7. Repeat the above six stages for the other two bacteria, *Serratia marcescens* and *Sarcina lutea*.

N.B. Incubate *Serratia* and *Sarcina* at 20°C.

8. Examine all the plates after incubation. Record the presence or absence of bacteria and make an estimate of the amount of growth produced.

Part 2

9. You will be provided with agar containing 1000 $\mu g/cm^3$ of streptomycin sulphate. From the results obtained in Part 1 above, choose the highest streptomycin concentration showing growth in all three organisms. Re-plate these on plates containing 1000 $\mu g/cm^3$ as follows.

10. Flame-sterilise an inoculating loop. Carefully pick up a loopful of bacteria from one of the results dishes obtained in Part 1. Streak out on a plate of nutrient agar plus 1000 $\mu g/cm^3$.

11. Repeat Stage 10 but using the original bacterial suspensions used in Stage 4. Incubate at the temperatures used in Part 1, for 48 h.

12. Describe the results obtained.

Part 3

13. Using the plates of bacteria prepared in Stages 9–12, subculture as you did previously in Stages 9–11, but onto ordinary nutrient agar, i.e. without streptomycin.

14. Again, describe the results of this third phase of the experiment.

QUESTIONS

1. In Part 1 of this experiment state on which dishes bacteria grew. Can you explain why this is?

2. Why are a range of streptomycin concentrations used to detect mutation?

3. Why are micro-organisms useful for demonstrating spontaneous mutations in a population?

4. Do the results suggest whether or not more than one gene is involved?

5. How do we know that the resistant forms are in fact mutants?

6. Do the mutant, streptomycin-resistant strains, now require the antibiotic as an intermediate metabolite?

7. What information does this experiment give us about the mechanism of evolution?

8. Have your results any significance with regard to the indiscriminate use of antibiotics in society?

EXPERIMENT 5.3

AN INVESTIGATION INTO BIOCHEMICAL MUTANTS IN THE BREAD MOULD—*NEUROSPORA CRASSA*

THEORY

Neurospora crassa is a flask fungus with a life cycle which is similar to that of *Sordaria* sp. (see Experiment 5.1). *N. crassa* has been used extensively in studying biochemical mutants, the aim being generally to elucidate details of particular biochemical pathways.

The first stage in such an investigation is to subject a wild type (i.e. non-mutant) *Neurospora* sp. to a mutagenic agent, e.g. X-rays, or ultraviolet light. The organism is then allowed to grow and produce conidia. These conidia may be mutant or normal. To find out, these conidia are crossed with a wild type of organism of the opposite mating strain.

This cross produces perithecia and asci. The ascospores produced, if we assume a mutation to have occurred, will be either mutant or normal, but there is no visible difference between them. They thus have to be removed and cultured individually.

The spores are removed from the ascus selected and are first kept at 60°C for 30 min. This 'heat shock' treatment breaks the spore dormancy. They are then put on a complete medium containing minerals, glucose, biotin, other vitamins, amino acids, purines and pyrimidines, any of which the organism may not be able to synthesise after mutation. All types of spores germinate and produce mycelia (i.e. both normal, wild and mutant types).

The mycelia so formed are then sub-cultured on a 'minimal' medium. This contains only the minimum requirement for the wild type *Neurospora* sp. It consists of: minerals, glucose, biotin and nitrate (as a nitrogen source). Any biochemical mutants fail to grow on this medium, thus showing a biochemical defect associated with the genotype of the spore (see *Flow Diagram 5.2*).

In order to determine the precise nature of this defect, the mutant is now cultured on a range of media. Initially the media used contain vitamins, or purines and pyrimidines, or amino acids, and the ability of the mutant to grow on the media is recorded (see *Table 5.2*).

Table 5.2

Minimal medium + vitamins	Minimal medium + purines and pyrimidines	Minimal medium + amino acids
No growth	No growth	Normal growth

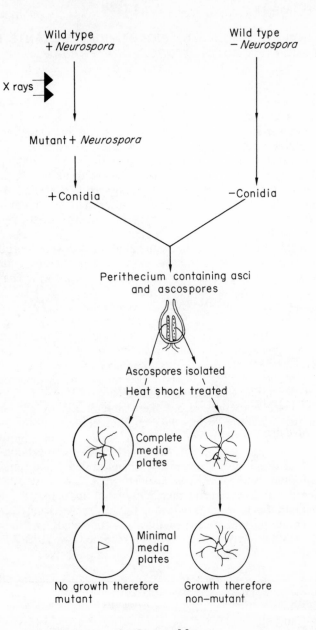

Flow Diagram 5.2

This suggests that the defect is an inability to synthesise a particular amino acid. Therefore, the next stage would be to culture the mutant on a variety of minimal media, each with only one amino acid added.

The following experiment uses an already established mutant and you will investigate the biochemical difference between a normal wild strain and the mutant strain.

PROCEDURE

1. Take four chemically clean 250 cm³ conical flasks. Make cotton wool plugs for these (see Experiment 1.4). Label the flasks with the Experiment No. and the numbers 1, 2, 3, and 4.

2. To flasks 1 and 2 add 50 cm³ of a minimal medium containing glucose, minerals, H_3PO_4/KOH buffer solution pH 6·0, biotin and sodium nitrate.

3. To flasks 3 and 4 add 50 cm³ of a medium containing glucose, minerals, H_3PO_4/KOH buffer solution, biotin and ammonium chloride.

4. Autoclave the flasks plus cotton wool plugs and media, at 10 lb/in² for 15 min.

5. Take a slope culture of *Neurospora crassa* wild strain and add 10 cm³ of sterile distilled water. Shake gently to produce a spore suspension. Inoculate flasks 1 and 3 with five drops of this suspension using a sterile Pasteur pipette. (See *Flow Diagram 5.3*.)

6. Take a slope culture of *N. crassa* mutant strain and add 10 cm³ sterile distilled water. Shake gently to produce a spore suspension. Inoculate flasks 2 and 4 with this suspension, again using five drops and a sterile Pasteur pipette.

7. Incubate at 35°C. After 2–3 days remove a small amount of the medium from all four flasks and test for nitrite as in Experiment 6.4, see page 124. (Room temperature can be used if necessary but use a longer period.)

8. Describe the appearances of the flasks, i.e. describe any growth that has occurred on them.

QUESTIONS

1. What can you say about the organism which cannot grow on the minimal medium?

2. Which nitrogen source is the mutant able to utilise?

3. Can you suggest the biochemical reactions which the mutant cannot carry out?

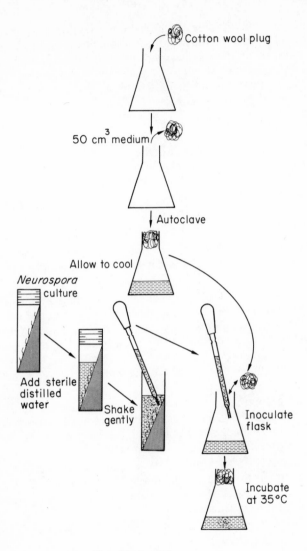

Flow Diagram 5.3

Table 5.3 RESULTS OF EXPERIMENT USING THREE MUTANTS OF *Neurospora* SP. SHOWING GROWTH ON THE MEDIA USED

			Growth of organism on:			
Mutant	Minimal medium	Minimal medium + amino acids	Minimal medium + all amino acids except arginine	Minimal medium + arginine	Minimal medium + citrulline	Minimal medium + ornithine
1	−	+	−	+	+	+
2	−	+	−	+	+	−
3	−	+	−	+	−	−

Key + growth
     − no growth

ANALYSIS OF A BIOSYNTHETIC PATHWAY

The study of mutants in experiments like the one you have just completed shows them to differ in their nutritional needs from the wild type *Neurospora*.

By further experiments using mutants of this type, the details of a particular biochemical pathway can be worked out. Consider the situation in *Neurospora* sp. (shown in *Table 5.3*) in which three related nutritional mutants are being tested. In this way it can be determined which specific amino acid the mutant was unable to synthesise.

Answer the following questions.

1. What do these results suggest with regard to the nutrition of all three mutants?
2. Which amino acids cannot be synthesised by any of the mutants?
3. Which substances can replace the amino acid for:
   > (*a*) mutant 1
   > (*b*) mutant 2
   > (*c*) mutant 3

   Does this suggest to you any way in which arginine, citrulline and ornithine can be placed in a reaction sequence? If so, what is the most logical sequence (give reasons for your answer).
4. Mutants are produced by alteration of the chemical nature of genes. How many genes do you think are involved in the reaction sequence?
5. How many enzymes do you think will be involved?
6. Does your answer to Questions 4 and 5 give you any clue as to the way in which genes may act to control such a reaction sequence?

6
Soil ecology

INTRODUCTION

Soil at first sight seems lifeless and even on close examination may reveal only a few macroscopic living things. Yet place this same soil under even a low-powered microscope and one begins to appreciate how inaccurate the term 'lifeless' is.

In the last 50 years the advent of refined biological, chemical, and physical techniques have led to a more thorough understanding of the dynamic nature of soil. We now know that it presents a constantly fluctuating environment in terms of mineral content, organic content, gases and solutions, and that a host of organisms exist in it. Certainly bacteria, fungi, actinomycetes, algae, protozoa, nematodes and insects, for example *Collembola*, all exist in the soil.

Many of these organisms are involved in the cyclic systems, for example Carbon and Nitrogen Cycles, etc., the continuous functioning of which are essential to the survival of many other living organisms. In addition, decomposing organisms remove litter and return valuable minerals etc., to the soil.

In this chapter you will investigate some of the inhabitants of the soil and estimate their numbers. You will also consider some aspects of the Nitrogen Cycle, the utilisation of cellulose and thus decomposition, and the ability of organisms to exist without oxygen.

EXPERIMENT 6.1

ESTIMATING THE NUMBERS OF MICRO-ORGANISMS IN SOILS

THEORY

One method of estimating the numbers of actinomycetes, bacteria and fungi in soils, is the dilution plate method. This employs the culture of a suspension of known dilution in a nutrient agar, and subsequent counting of the colonies which develop. As the dilution factor is known, the number of colonies per gramme of soil can be calculated. This method can be used to compare the numbers of organisms in different soils.

In this experiment you will investigate the number of bacteria and actinomycetes, filamentous fungi and yeasts in two different soils.

PROCEDURE

You will be provided with samples of garden and forest soils which have been collected by pushing a sterile tube into the soil.

1. Using a clean balance weigh out 10 g of garden soil into a sterile beaker.
2. Put this soil into an already sterilised 250 cm^3 conical flask containing 90 cm^3 of sterile distilled water, and plugged with a cotton wool plug. Shake the flask thoroughly to disperse the organisms.
3. Obtain a series of six sterile bacteriological tubes plus caps, each containing 9 cm^3 of sterile distilled water.
4. Take a 1 cm^3 sterile pipette and take up some of the soil suspension from the 250 cm^3 flask. MAKE SURE YOU SHAKE THE FLASK BEFORE YOU PIPETTE YOUR SAMPLE. Put 1 cm^3 of this suspension into the first bacteriological tube. Discard the pipette. Shake the tube to ensure even dispersion of the 1 cm^3 inoculum. Label this tube 1.
5. Take another 1 cm^3 sterile pipette and take up some of the suspension you have just prepared in tube 1. Put 1 cm^3 of this suspension into the second bacteriological tube. Discard the pipette and shake the tube. Label this tube 2.

In Stage 4, you have prepared a 1/100th dilution, as the original suspension was a 1/10th dilution. Therefore in Stage 5 you have prepared a 1/1000th (10^3) dilution.

6. Continue the procedure described in Stage 5 to prepare 10^4, 10^5, 10^6, and 10^7th dilutions in the last four bacteriological tubes. USE A FRESH STERILE PIPETTE EACH TIME.
7. Take two sterile Petri dishes. Using a sterile pipette take up some of the 10^7th dilution suspension and pipette 1 cm^3 of this suspension into each of the two Petri dishes.

   Pour 15 cm^3 of sterile malt extract agar (containing rose bengal and streptomycin sulphate) taken from the 45°C water-bath, into the first of these dishes. Mix with the soil suspension by gently swirling the dish.

   Into the second dish pour 15 cm^3 of sterile nutrient agar at 45°C. Mix by swirling gently.

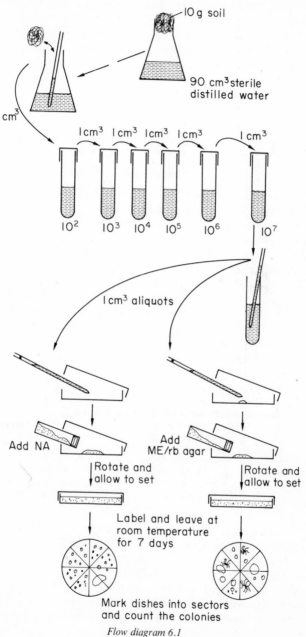

Flow diagram 6.1

Label both dishes with the Experiment No., dilution factor and Group No.

8. Repeat Stage 7 for the remaining five dilutions.

N.B. If you start at the greatest, i.e. 10^7 dilution and work upwards you can use the same sterile 1 cm³ pipette to remove all the 1 cm³ amounts.

9. Leave the twelve Petri dishes at room temperature for 5–7 days.
10. Repeat the procedure using a forest soil.
11. Measure the pH of the soils using a pH meter or soil indicator kit.

Estimating the numbers of bacteria, fungi, actinomycetes and yeasts

The media used in the above experiment are to a large extent selective. Thus the malt extract agar plates will tend to grow fungi, whereas the NA plates will grow mainly bacteria.

Your dilution plates for each soil will show a series in terms of numbers of colonies. Thus the lowest dilution, i.e. 10^2, will show the few colonies or none at all. Select those plates with 30–300 colonies.

12. Using a grease pencil (or felt-tip pen), divide the backs of the dishes chosen into eight equal sectors. Hold the plate over a microscope lamp or up to the light and count looking through the back of the plate. Count the colonies per sector and make a mark over each colony as you count it to avoid counting a colony twice.

13.

Malt agar plates pH 5·4	*Nutrient agar plates* pH 7·4
Count the number of fungal colonies. Any smooth colonies appearing will either be bacteria or yeasts. Check by making a wet mount of these colonies and examining under the high power of the microscope. Yeasts will appear as large rounded cells. Count the number of yeasts—ignore bacteria	Count the number of bacterial colonies. Count the number of curly colonies, or large colonies with a chalky consistency. These are actinomycetes (filamentous bacteria). Ignore any fungal colonies on the plates

14. Calculate the numbers of bacteria, filamentous fungi, yeasts, and actinomycetes per gramme of soil, for both soils.

QUESTIONS

1. Why do you think so many organisms are present in soil?
2. Are all the organisms isolated from soil necessarily active in soil environments? If not, where do you think they come from?
3. Are there any major differences between the soils you have

studied? If differences do exist, can you give any reasons for them?

4. Why do you think the media used are more or less selective for fungi or bacteria?

EXPERIMENT 6.2

CELLULOSE DECOMPOSITION BY SOIL MICRO-ORGANISMS

THEORY

The most important source of organic material added to the soil is dead plant tissue. The materials which constitute the protoplasm and storage materials are rapidly decomposed because they are available to a wide range of organisms. Skeletal framework materials, however, such as cellulose and lignin, are less readily decomposed.

In this experiment you will investigate the decomposition of cellulose strips added to soil.

PROCEDURE

1. You will be provided with three types of soil, namely: forest soil, garden soil and sterile soil (oven-sterilised at $160°C$ for $1\frac{1}{2}$ h).
2. Obtain two sterile Petri dishes. Into both dishes put enough forest soil to cover the bottom of the dish.
3. You will be provided with a Petri dish containing strips of sterile Cellophane. Using sterile forceps, remove four strips and place them on the surface of the forest soil in one of the Petri dishes. Cover the strips with more forest soil. Repeat this with the other Petri dish containing forest soil. Label one dish Experiment No. 6.2, Group No. . . . , forest soil A, and the other dish Experiment No. 6.2, Group No. . . . , forest soil B.
4. Repeat Stages 2 and 3 using garden soil. Label accordingly.
5. Repeat Stages 2 and 3 using sterile soil. Label accordingly.
6. Divide the dishes into two sets, dishes A and dishes B. Moisten the soil in set A using distilled water and a sterile pipette (graduated). Note the volume used in the first dish and add the same volume to each successive dish. Moisten the soil in set B using a solution consisting of 0.1% ammonium sulphate and 0.1% potassium hydrogen phosphate. Again use a sterile graduated pipette and use the same volumes as were applied to set A.
7. Put on one side a few Cellophane strips in a sterile sealed container.
8. Leave the dishes at room temperature for 1 week.

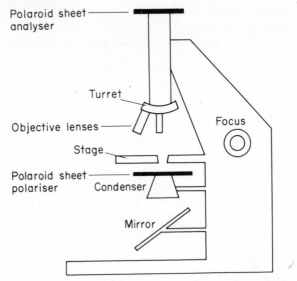

Figure 6.1 *To show the arrangement of Polaroid sheets on the microscope*

ASSESSMENT OF RESULTS

9. You will examine the Cellophane strips using polarised light. Set up your microscope as shown in *Figure 6.1*, with a Polaroid sheet between the condenser and the stage, and another Polaroid sheet on top of the eyepiece lens.
10. Using the low-power objective lens, look down the microscope and rotate the ANALYSER sheet, i.e. the Polaroid sheet on top of the eyepiece, until the field of view is as dark as you can make it. Leave the two sheets exactly where they are.
11. Now mount a strip of Cellophane (saved from last week—Stage 7) in water and place it on the stage. Observe that the dark field may change to a light one. This is because the Cellophane has caused a bending of the light, i.e. the Cellophane is optically active. If the field produced is not a light field, rotate the Cellophane to produce a light field. Now rotate the analyser to produce as dark a field as possible. This will probably be dark blue. Leave the Polaroid sheets exactly where they are.
12. Take one of the dishes containing Cellophane strips in soil. Use a pair of forceps to carefully remove one of the strips. Shake it GENTLY in a beaker of distilled water to remove soil particles, and mount it in water on a clean glass slide.

    Examine the strip under the microscope with the polariser and analyser in the positions obtained in Stage 11.

Unchanged Cellophane will appear as dark blue. Where partial decomposition of the Cellophane has occurred, brown–red areas will show. Total decomposition is shown by either yellow or white areas (depending on the light source).

13. Construct a table as below (*Table 6.1*) and fill in the results for the various soil situations. Use the following scale:

$$0 = \text{no change}$$
$$+ = \text{a little decomposition, etc.}$$
$$+ + + + = \text{considerable or almost total decomposition}$$

Table 6.1 ASSESSMENT OF DECOMPOSITION OF CELLOPHANE

Soil type		Moistened with distilled water	Moistened with nutrient solution
Sterile	1		
	2		
	3		
	4		
Average			
Garden	1		
	2		
	3		
	4		
Average			
Forest	1		
	2		
	3		
	4		
Average			

14. Having assessed decomposition choose one Cellophane film which shows a lot of decomposition and boil it in a few cubic centimetres of Benedict's solution. Do the same with one of the Cellophane strips which has not been in soil. Note any changes which occur in either strip.

15. Again choose a Cellophane strip which shows considerable decomposition. Place it in phenolic rose bengal on a slide. Heat gently for a few minutes BUT DO NOT BOIL. Dehydrate in 100% alcohol in a watch glass for 5 min. Transfer to xylene for 2–3 min and mount in Canada balsam. Examine for fungal hyphae and bacteria—living cytoplasm stains red.

QUESTIONS

1. How far does the experiment show that micro-organisms are responsible for Cellophane decomposition in the soil?
2. How do you think the process of Cellophane decomposition is brought about by soil micro-organisms?
3. What do you think (from the results of your Benedict tests) the products of decomposition of Cellophane might be?
4. What happens to the products of Cellophane decomposition?
5. Are there any differences in the rate of decomposition between the garden soil and the forest soil? Attempt to explain any differences you notice.
6. What is the effect of moistening the soil with nutrient solution? Explain any differences you notice.
7. Can you explain why polarised-light microscopy is used to show the difference between digested and undigested Cellophane?

EXPERIMENT 6.3

AMMONIFICATION IN SOILS

THEORY

During decomposition of dead plant and animal material, any organic nitrogen present is converted to inorganic compounds which are then available for higher plant growth. This process may be called **mineralisation**, and the first stages of this process involve the release of ammonia as a result of the decomposition of organic nitrogen by micro-organisms. This release of ammonia is referred to as **ammonification.**

In alkaline soils to which large quantities of either manure or nitrogenous fertilisers have been added, ammonia may be released to the atmosphere as gas and therefore lost to plants. Under more normal soil conditions, however, the ammonia is immediately adsorbed on the surface of colloidal soil particles where further chemical transformations occur (see Experiment 6.4).

In this experiment you will investigate the production of ammonia by soil organisms.

PROCEDURE

1. Take 12 sterile 250 cm^3 conical flasks with small beakers as covers.
2. Label the flasks 1–12.
3. Into each of flasks 1–8, weigh out 50 g of the fresh soil provided.

4. Into each of flasks 9–12, weigh out 50 g of the sterile soil provided.

5. Take some legume meal (e.g. ground dried peas or beans) and add 0·5 g of this to flasks 1–4 (inclusive) and 9–12 (inclusive).

6. Flame-sterilise a spatula, allow to cool and then mix the soil and legume meal in flask 1. Repeat for the other seven flasks prepared in Stage 5.

7. Moisten the soils with sterile distilled water.

N.B. DO NOT add too much water or you will create anaerobic conditions.

8. Incubate all your flasks together at room temperature.

ASSESSMENT

9. Analyse one flask of non-sterile soil, for example flask 5, one flask of non-sterile soil+legume meal, for example flask 1, and one flask of sterile soil, for example flask 9, as follows:

   (a) Pour in 100 cm³ of 1N NaCl solution into the flask. Stopper, using a rubber bung, and shake intermittently for 30 min. (Use a mechanical wrist-action shaker if available.)

   (b) Filter the shaken soil suspension through a Buchner funnel using a coarse filter paper or glass wool—or allow to stand and decant the liquid.

   (c) Test this filtrate for ammonium ions by adding 0·5 cm³ Nessler's reagent to 5·0 cm³ of the extract. Indicate the intensity of the colour produced using a qualitative scale, for example

$$0 \longrightarrow + + + +$$

   no coloration   intense orange precipitate

10. Repeat the above test after 3 days, 7 days and 14 days and tabulate your results.

QUESTIONS

1. Why is legume meal used in the experiment rather than other plant material?

2. What imporant nitrogen containing material is present in the legume meal which may give rise to the ammonia on decomposition?

3. What does your experiment suggest with regard to the nitrogen status of an unamended soil?

4. What changes occurred in the amended flasks during the experimental period? Attempt to explain the changes observed.

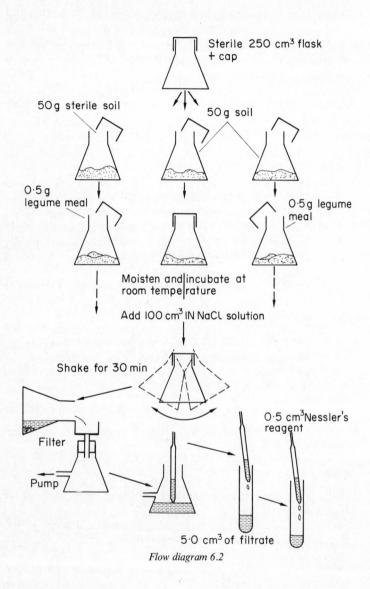

Sterile 250 cm³ flask + cap

50 g sterile soil

50 g soil

0·5 g legume meal

0·5 g legume meal

Moisten and incubate at room temperature

Add 100 cm³ 1N NaCl solution

Shake for 30 min

Filter

Pump

0·5 cm³ Nessler's reagent

5·0 cm³ of filtrate

Flow diagram 6.2

5. Why is an extraction with NaCl used, before testing for ammonia?
6. Why is sterile soil used as part of the experiment?

EXPERIMENT 6.4

NITRIFICATION BY SOIL ORGANISMS

THEORY

A source of nitrogen is essential to all organisms for the formation of proteins, etc., and although the atmosphere contains some 80% nitrogen gas, this is not available to the majority of organisms. For higher plants living in soil, nitrate is the most easily available form of nitrogen.

Micro-organisms in the soil cause the breakdown of proteins (from dead organic remains) with the subsequent release of ammonia. This ammonia is used by other micro-organisms to form the nitrate so essential to higher plants. This process is called nitrification and is carried out by the soil bacteria *Nitrosomonas* sp. and *Nitrobacter* species.

In this experiment you will attempt to demonstrate nitrification in soil.

PROCEDURE

1. Take four sterile 250 cm³ flasks with cotton wool plugs.
2. Label the flasks 1–4.
3. Aseptically transfer 50 cm³ of a medium plus 0·5% ammonium sulphate to each of flasks 1 and 2.
4. Similarly add 50 cm³ of a medium+0·5% sodium nitrite to flasks 3 and 4.
5. Take some garden soil and weigh out 1·0 g. Add this to flask 1. Repeat for flask 3.
6. Weigh out two separate 1·0 g amounts of the same soil and autoclave them at 15 lb/in² for 15 min. Remove and allow to cool. Add one to flask 2 and the other to flask 4.
7. Shake the flasks to ensure an even dispersion of the soil.
8. Test the original solutions, i.e. the 0·5% ammonium sulphate, the 0·5% sodium nitrite, and a 0·5% solution of potassium nitrate, as follows:

For nitrate: Add one drop of the solution under test to one drop of diphenylamine reagent. Carry out the test on a white tile. A blue colour indicates the presence of nitrate.

For ammonia: Add one drop of the test solution to one drop of Nessler's reagent. Again do this on a white tile. A yellow-orange colour indicates the presence of ammonia.

For nitrite: Take 3 cm³ of sulphanilic acid/acetic acid reagent and mix it with 3 cm³ of naphthalene/acetic acid reagent. Then add one drop of this mixture to a drop of the test solution. Do this on a white tile. A red colour is positive for nitrite.

Use the results of these tests as standards.

9. Now test the solutions in the flasks for ammonia, nitrite and nitrate.
10. Leave all flasks to incubate at room temperature for 2 weeks.

TESTING FOR RESULTS

11. Test the flasks for the above compounds after 5 days incubation.
12. Repeat the testing after 10 days and 15 days. You may wish to retain some of the original solutions added to the flasks in order to perform comparative tests at these times, and to use 0·5% KNO_3.
13. Tabulate your results showing the intensity of colour developed with the tests as follows:

$$+ \longrightarrow + + + + +$$

very weak colour   colour fully developed (i.e. that produced by a 1·0% solution)

14. Explain your results.

QUESTIONS

1. What evidence does the experiment give that living organisms are responsible for the conversion of ammonium to nitrate in the soil, rather than any other process?
2. Is there any direct evidence from your experiment, for the idea suggested above, that bacteria, rather than any other group of organisms, are responsible for nitrification?
3. Conversion of $NH_4^+ \longrightarrow NO_3^-$ is normally considered to be a two-stage process in which:

   (a) Ammonia is oxidised to nitrite by *Nitrosomonas* sp.

   $$NH_4^+ \longrightarrow NO_2^-$$

   and

   (b) Nitrite is further oxidised to nitrate by *Nitrobacter* sp.

   $$NO_2^- \longrightarrow NO_3^-$$

However, nitrite is toxic to plants and is seldom found free in soils. What does this suggest with regard to the distribution of the two micro-organisms?

4. *Nitrobacter* and *Nitrosomonas* are both autotrophic, i.e. they can synthesise organic materials using carbon dioxide as the sole source of carbon. These bacteria, however, contain no photosynthetic pigments. Where do you think they obtain the energy for conversion of carbon dioxide to organic carbon compounds?

5. Although nitrate is the source of nitrogen most easily available to higher plants it is also the form most easily lost from soils. How does this occur?

EXPERIMENT 6.5

CULTURING ANAEROBIC ORGANISMS FROM THE SOIL

THEORY

Soil organisms can be divided into three groups, in relation to the way in which they respond to the presence or absence of oxygen in their environments. These three are:

(*a*) Those that grow only in the presence of oxygen—**the obligate aerobes**.

(*b*) Those that grow only in the absence of oxygen—**the obligate anaerobes**. In fact their growth is often inhibited by the presence of oxygen.

(*c*) Those that can grow whether oxygen is present or absent—**the facultative anaerobes**.

Normal culture methods employing Petri dishes or flasks plugged with cotton wool allow the organism being cultured contact with the atmospheric oxygen. Thus only obligate aerobes or facultative anaerobes will grow in these cultures. Obligate anaerobes, however, form an important part of the soil microflora, but they can only be cultured in an oxygen-free environment.

In this experiment you will aim to produce an oxygen-free environment and to culture anaerobic soil organisms.

PROCEDURE

1. Take a thoroughly clean desiccator and check that the ground glass edges are greased with Vaseline to make a proper air-tight seal when the lid is closed.

2. Prepare a soil sample as follows:

(*a*) Weigh out 10 g of soil and add 90 cm³ of sterile distilled water, in a sterile 250 cm³ flask.

(*b*) Shake to make an even dispersion.

(*c*) Using a sterile pipette remove some of this suspension and pipette 1 cm³ aliquots into six sterile Petri dishes.

3. To each Petri dish add 15 cm³ of nutrient agar which has been maintained at 45°C. Gently rotate the dish to mix the agar and inoculum, and allow to set.

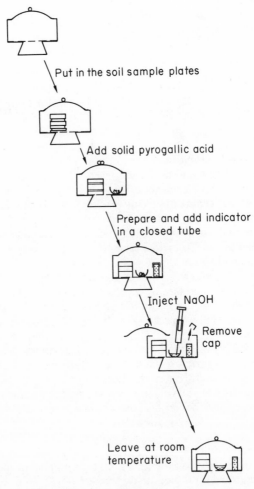

Put in the soil sample plates

Add solid pyrogallic acid

Prepare and add indicator in a closed tube

Inject NaOH

Remove cap

Leave at room temperature

Flow diagram 6.3

4. Label all the dishes with the Experiment Number and your Group Number. Stand three of the Petri dishes in the desiccator. The other three act as controls and remain outside in an aerobic environment.
5. Weigh out approximately 10 g of pyrogallic acid and put this into an evaporating dish. Place the dish into the desiccator. (See *Flow Diagram 6.3.*)
6. You will need an indicator to show whether or not anaerobic conditions have been obtained. Prepare about 15 cm^3 of the following indicator:

> Mix together 5 cm^3 amounts of these three reagents:
> 0·02% NaOH
> 0·015% methylene blue
> 6·0% glucose

Boil the mixture until it is completely colourless and pour it into a specimen tube with a tight-fitting lid, for example plastic snap-top lid. Fill the tube as full as you can to avoid the indicator coming into contact with atmospheric oxygen. Place the sealed tube plus indicator in the desiccator and allow it to cool.
7. Take a syringe and needle and fill it with 20 cm^3 of 1 N sodium hydroxide.
CARRY OUT THE FOLLOWING STAGES QUICKLY.
8. Open the desiccator and remove the lid of the indicator tube. Inject the 20 cm^3 of 1 N NaOH into the dish containing the pyrogallic acid, thus producing a solution of alkaline pyrogallol. Close the lid of the desiccator and ensure an air-tight seal by rotating the lid a small amount.
The alkaline pyrogallol absorbs the oxygen from the air, and except for the surface layer, the indicator should remain colourless. An unduly large amount of blue colour will indicate the presence of oxygen. Compare this with a similar tube of indicator left outside the desiccator.
9. Check after 24 and 48 h that an anaerobic atmosphere has been produced. If not repeat Stages 5–8 inclusive.
10. Leave for 1 week at room temperature.
11. Examine the desiccator. Open the desiccator and compare the growth on Petri dishes from inside the desiccator with those outside. Describe your results.

QUESTIONS

1. What causes the noticeable smell when you open the desiccator?
2. In a normal well aerated soil, do aerobic or anaerobic organisms predominate?

3. In which natural environments do you expect to find large numbers of anaerobic organisms.

4. Under anaerobic soil conditions, organisms may still require substances containing oxygen to act as hydrogen acceptors in order to oxidise organic substrates. Which inorganic substances, present in soil, may fulfil this role? What do you think the products of such reduction reactions might be?

5. What role(s) may anaerobic organisms play in natural environments?

EXPERIMENT 6.6

CULTURING SOIL PROTOZOA

THEORY

The micro-organisms present in the soil belong to many different groups, some of which have been seen in Experiment 6.1. Other groups of micro-organisms occur however, and particularly abundant are the Protozoa.

The most commonly occurring types are amoebae and flagellates, but ciliates may also be present in large numbers. In the soil many of the amoeboid forms—the Rhizopoda have a chitinous shell and are called thecate or testaceous rhizopods.

Normally protozoans are active in the water films around soil particles, but under dry conditions they exist as cysts or resting cells.

PROCEDURE

You will be provided with two soil samples, one of garden soil, the other of acid (peaty) soil.

1. Carefully weigh out 10 g of garden soil into a sterile 250 cm³ flask. Add 100 cm³ of sterile distilled water. Shake the flask thoroughly to make an even soil suspension.

2. Take two sterile Petri dishes. Take 2×15 cm³ amounts of sterile sodium chloride agar from the 45°C water-bath. Pour two NaCl agar plates. Take 10 of the sterile anodised aluminium or glass rings provided. USING STERILE FORCEPS place five of the rings in each plate whilst the agar is still molten—see *Flow Diagram 6.4.*
   Allow the plates to set so that the rings are set into place.

3. Label your plates NaCl agar, Experiment 6.6, Group No. . . . , plates gs 1 and gs control.

4. Now take a 48 h broth suspension of the bacterium *Aerobacter aerogenes.*

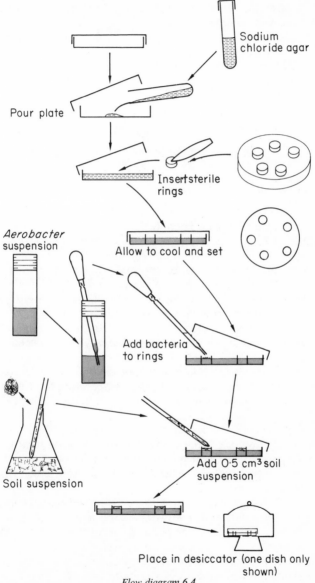

Sodium chloride agar

Pour plate

Insert sterile rings

Aerobacter suspension

Allow to cool and set

Add bacteria to rings

Soil suspension

Add 0·5 cm³ soil suspension

Place in desiccator (one dish only shown)

Flow diagram 6.4

5. Take up some of the suspension into a Pasteur pipette.
Using an aseptic technique place enough of the suspension dropwise into each ring of plate gs 1, so as to cover the bottom of the well (see *Figure 6.2*).

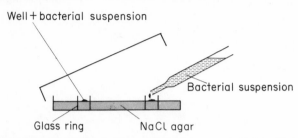

Figure 6.2 *To show Petri dish + rings and bacterial suspension being added*

Repeat this for plate gs control.
6. Reshake the suspension of garden soil and take up some of the suspension into a sterile 1 cm³ pipette.
7. Add about 0·5 cm³ (or less or more, depending on the size of your rings) of the suspension to each of the 'wells' in plate gs 1.
8. Now add 0·5 cm³ (or the amount used in Stage 7) of sterile distilled water using ANOTHER sterile 1 cm³ pipette, to the 'wells' in plate gs control.
9. Repeat Stages 1–8 using the acid (peaty) soil, labelling your plates as 1, and as control.
10. Now place all four plates into a dessicator which has water in the bottom and incubate at room temperature for 1 week.

ASSESSMENT

11. After 1 week, take your four plates out of the desiccator.
12. Using a sterile Pasteur pipette take a small drop of culture from one ring of plate gs 1. Mark the dish to show which ring was sampled.
13. Mount the drop, together with a drop of neutral red stain, on a clean glass slide and cover with a cover slip.
Examine under the microscope for Protozoa and record your findings.
14. Repeat with the other three plates.
15. Repeat the assessment daily, rotating to each of the rings in turn.
(OR. Assess all the rings at weekly intervals.)

QUESTIONS

1. Are Protozoa present in soils?
   If so, from your results can you suggest which groups of Protozoa are most common?
2. Are there any differences between the Protozoan populations of garden, and acid (peaty) soils?
3. What is the purpose of the *Aerobacter aerogenes*?
4. Why is it necessary to put the plates in a desiccator with water in the bottom?

EXPERIMENT 6.7

CULTURING SOIL ALGAE

THEORY

Algae are one group of autotrophic organisms present in soil. They have photosynthetic pigments, for example chlorophylls and accessory pigments, which enable them to utilise the energy from sunlight to produce organic substances using carbon dioxide as a raw material. Some of these algae are also capable of fixing atmospheric nitrogen, i.e. incorporating nitrogen gas from the atmosphere into organic nitrogen compounds.

The most important algal groups represented in soils are: the blue-green algae, the green algae, the yellow-green algae and the diatoms (see *Figures 6.3* (a), (b) and (c) on pages 132 and 133).

Soil algae are usually difficult to study by direct examination of soil. However, by supplying them with nutrient materials in laboratory culture we can increase their numbers sufficiently for studies to be made. We call this process enrichment culture.

PROCEDURE

1. Collect some surface soil, preferably from the surface of a damp, exposed, relatively undisturbed soil area.
2. Collect a sterile 250 cm³ flask which contains 50 cm³ of a sterile medium called Bristol's solution, and which is plugged with a cotton wool plug.
3. Weigh out 1 g of your soil and using aseptic technique transfer it into the flask. Shake thoroughly to disperse the soil evenly.
   Bristol's solution is a medium containing the necessary enrichment nutrients for algal culture, for example sodium nitrate, calcium chloride, magnesium sulphate, dipotassium hydrogen phosphate, potassium dihydrogen phosphate, sodium chloride and ferrous chloride.

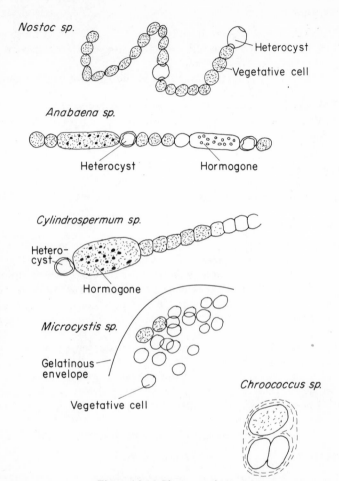

Figure 6.3 (a) *Blue-green algae*

4. Take another flask plus plug and containing 50 cm³ of sterile Bristol's solution, but without the sodium nitrate. Inoculate this with one gramme of soil as above. This medium will be selective for those organisms that can fix atmospheric nitrogen.
5. Label your flasks with the experiment number, your group number and the symbols CM for complete medium and CM—N for the one less sodium nitrate.
6. Leave your flasks under a growth promoting light source, or in strong daylight (BUT NOT in direct sunlight) for several weeks.

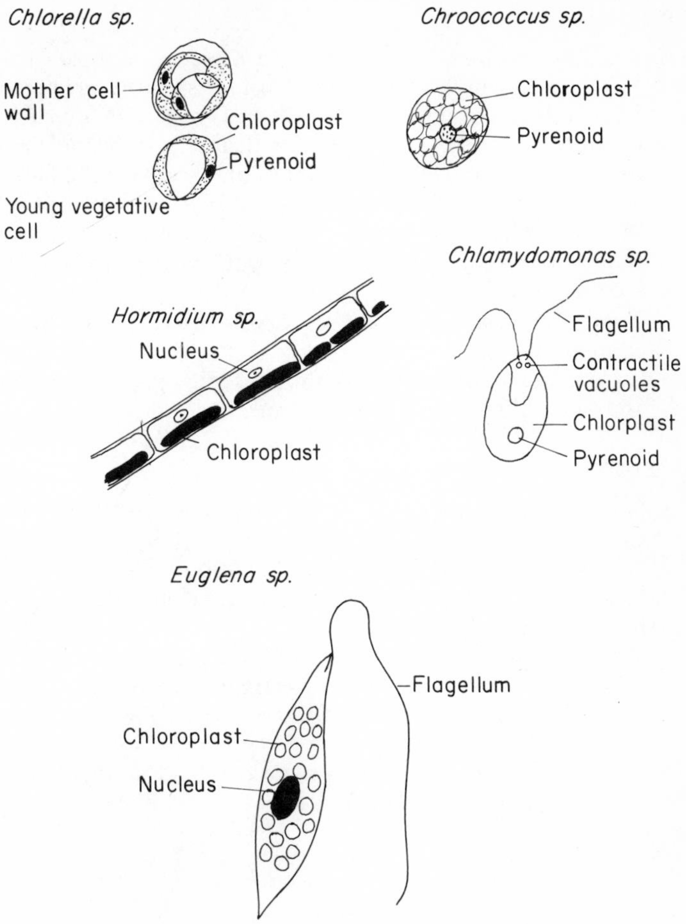

Chlorella sp.

Mother cell wall
Chloroplast
Pyrenoid
Young vegetative cell

Chroococcus sp.

Chloroplast
Pyrenoid

Hormidium sp.

Nucleus
Chloroplast

Chlamydomonas sp.

Flagellum
Contractile vacuoles
Chlorplast
Pyrenoid

Euglena sp.

Flagellum
Chloroplast
Nucleus

Figure 6.3 (b) *Green algae*

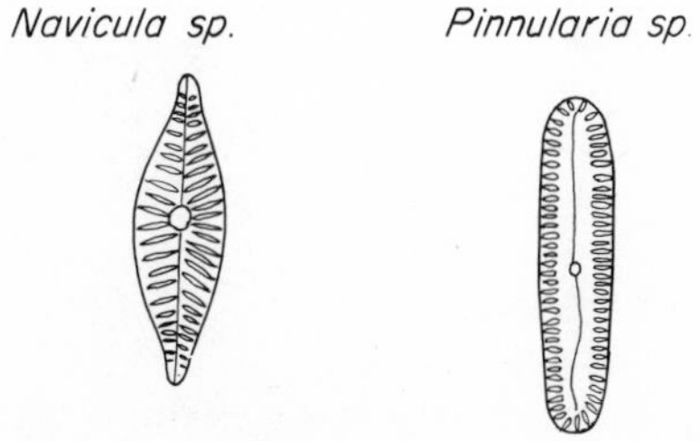

Navicula sp.

Pinnularia sp.

Figure 6.3 (c) *Diatoms*

ASSESSMENT

7. Examine the flasks weekly for the presence of algal growth. Take a small sample of the medium using a sterile Pasteur pipette. Place a drop on a clean glass slide and cover with a cover slip. Examine under the microscope for algal cells. You may find the following types: non-motile unicellular forms, flagellate unicellular forms, and filamentous algae. Also flat plates of cells together. Refer to *Figures 6.3* (a), (b) and (c) which show some of the more common types of algae you are likely to find. Try to identify as many types as you can.

   Note: The flasks containing the CM—N medium are likely to contain blue-green algae.

   Compare your results with those of other groups in the class and compile a list of all species found.

QUESTIONS

1. How would you expect algae to be distributed vertically in the soil?
2. Why would you expect such a distribution.
3. Under what ecological conditions would you expect soil algae to occur most abundantly?
4. Why is the medium used selective for algae?
5. What are the possible disadvantages of the enrichment technique in giving an overall picture of the types of algae present in the soil?
6. How would you attempt to isolate one kind of alga from the mixed population in your flasks?

7

Ecology

INTRODUCTION

The growth of micro-organisms in any habitat will be influenced by environmental variables such as pH, temperature, and osmotic pressure. These environmental effects will vary for different habitats, so that micro-organisms growing in water, or milk, or on food materials may be different. Some aspects of these variables are considered in the first three experiments of this chapter.

Water, milk, and food materials are microbial environments with which man is in direct contact, and all three of these may carry pathogenic organisms, or the toxic by-products of microbial metabolism. In addition, other micro-organisms found in milk and food materials utilise them for growth and by so doing cause food spoilage, i.e. render the foods unfit for human consumption. Experiments 7.4, 7.5, and 7.6 are concerned with these three complex natural habitats and some aspects of their microbiology.

EXPERIMENT 7.1

TEMPERATURE AND THE GROWTH OF MICRO-ORGANISMS

THEORY

You have already seen in previous experiments that micro-organisms can grow in a variety of habitats. One factor involved in the ability of an organism to colonise a particular environment is temperature. This factor limits the range of environments however, as micro-organisms can only grow within a temperature range of approximately —5°C to +95°C.

Micro-organisms can be grouped according to the temperature range over which they can grow, and to their optimum growth temperatures, into **thermophiles**, **mesophiles** and **psychrophiles** (see *Figure 7.1*).

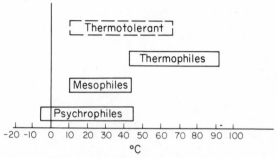

Figure 7.1 *To show the relationship between temperature and the growth of micro-organisms*

Each individual micro-organism, however, will have its own specific maximum, minimum and optimum temperatures, and these may not fit the defined limits of the groups shown above. Thus these divisions are arbitrary ones and some organisms will overlap the boundaries shown, for example thermotolerant organisms.

Although active growth may be defined to a rather limited temperature range, the organism will often survive over a much wider range in temperature. This is often associated with the production of special temperature-resistant bodies, for example bacterial endospores.

In the following experiment you will investigate the effect of temperature on the growth of a number of micro-organisms.

PROCEDURE

You will be provided with 24 h nutrient broth cultures of the following bacteria:

> *Pseudomonas fluorescens*
> *Escherichia coli*
> *Serratia marcescens*
> *Bacillus stearothermophilus*

You will be provided with slope cultures of the following fungi:

> *Humicola lanuginosa*
> *Byssochlamys fulva*
> *Mucor plumbeus*

1. Take a 24 h broth culture of *P. fluorescens*. Take also a bacteriological tube plus cap containing 10 cm³ sterile nutrient broth.
2. Flame-sterilise an inoculating loop and allow to cool.
3. Using aseptic technique pick up a loopful of the *P. fluorescens* culture and transfer it to your tube of broth. Label this tube 1 P.f.
4. Repeat Stages 1–3 to produce three more tubes, inoculated with *P. fluorescens*. Label these tubes 2 P.f., 3 P.f., etc.
   Note: This assumes that you have four temperature situations. (Check this with your supervisor and if necessary prepare more tubes.)
5. Now repeat Stages 1–4 for the remaining three bacteria.
6. Carefully label all your tubes, with Experiment No., Group No., and organism.
7. Divide the tubes into sets of four, each set containing one tube of each bacterium. Place a set in each of the following incubation situations:

Domestic refrigerator	circa 5°C
On the laboratory bench	
(room temperature)	15–20°C
Incubator 1	37°C
Incubator 2	55°C

8. Take the slope cultures of fungi. Take 12 sterile Petri dishes and 12 lots of sterile potato dextrose agar (PDA) at 45 °C.
   Pour 12 PDA plates (this again assumes four incubation temperatures), allow to set and label PDA.
9. Flame-sterilise a scalpel and allow to cool.
10. Using aseptic technique cut a small triangle of mycelium + agar from the *Humicola lanuginosa* culture. Place it face down on the centre of the first PDA plate.
11. Repeat Stages 9 and 10 and inoculate three more plates. Label them all with Experiment No., Group No., and the symbol *H. lan.*
12. Now repeat Stages 9–11 for the other two fungi and label with appropriate symbols (*B. fuv.* and *M. pl.*).
13. Place your tubes and plates in the various incubation situations and leave for 1 week.
14. Examine the tubes and plates at the end of the incubation period and assess results as follows.

ASSESSMENT

15. Collect together all the tubes and plates for each of the organisms used. Examine them and carefully assess the amount of growth.

Tabulate your results as shown in *Table 7.1*, using an arbitrary scale ranging from

0=no growth to $+$=little growth to $++++$=maximum growth

Table 7.1 RESULTS TABLE

Name of organism	Temperature			
	5°C	Room, °C	37°C	55°C

Comment on the pattern of growth, etc., and answer the following questions.

QUESTIONS

1. Which of the organisms used in the experiment are (*a*) psychrophilic, (*b*) mesophilic and (*c*) thermophilic?
2. What effect does temperature have on the production of pigment by *Serratia marcescens*? Can you give possible reasons for this effect?
3. What are the approximate maximum, minimum and optimum temperatures for growth of these organisms?
4. On the basis of your experimental results what do you think the natural habitats of these organisms might be?
5. Can you attempt to explain why some organisms are able to grow at relatively high temperatures, i.e. are thermophilic?
6. Why is the growth of micro-organisms limited to a rather narrow temperature range?
7. How is our knowledge of the relationship between temperature and the growth of micro-organisms applied to methods of food preservation such as refrigeration, pasteurisation, canning and sterilisation?

EXPERIMENT 7.2

THE EFFECT OF pH ON THE GROWTH OF MICRO-ORGANISMS

THEORY

In pure water at 25°C dissociation of the water molecules into hydrogen ions and hydroxyl ions occurs:

$$H_2O \rightleftharpoons H^+ + OH^-$$

This occurs to the extent that there are 10^{-7} g ions/litre of H^+ and 10^{-7} g ions/litre of OH^-, i.e. the solution is neutral.

Acid solutions contain more hydrogen ions (H^+) than hydroxyl (OH^-) ions but the total number of ions present is still 10^{-14}. Thus an acid solution containing 10^{-5} H^+ g ions/litre will also contain 10^{-9} OH^- g ions/litre. Alkaline solutions similarly have a predominance of OH^- ions and subsequently fewer H^+ ions.

Since it is rather a nuisance to keep writing down amounts like 10^{-9} g ions/litre, the pH scale was introduced as a more convenient term of reference. By taking log_{10} of the reciprocal of the value we obtain of the number of ions in a solution, we convert the unwieldy phrase above to simple whole numbers. The scale only relates to H^+ ions—hence the term pH. Thus a solution containing 10^{-14} g ions H^+/litre will be pH 4·0 (acid), and one containing 10^{-8} g ions H^+/litre will be pH 8·0 (alkaline).

The pH scale will then run from 0–14, because the maximum possible number of H^+ ions is 10^0 and the minimum number is 10^{+14}. The neutral point where numbers of H^+ = numbers of OH^- is pH 7·0. This can be expressed:

$$0 \text{———————} 7 \text{———————} 14$$
maximum possible   neutral   maximum possible
acidity   alkalinity

Note that every change of a pH unit, i.e. 2·0–3·0, represents a tenfold increase (or decrease) in the number of hydrogen ions present in the solution.

pH is another environmental variable which influences the growth of micro-organisms. In general, micro-organisms are limited by extremes of pH, but can nevertheless grow over a fairly wide pH range. They certainly grow best, however, at the pH of their natural environment, and this pH is often the **optimum pH** for a particular organism, i.e. the pH at which growth is most rapid. The same organism will have a maximum pH (high H^+ concentration) and a minimum pH (low H^+ concentration) at which no growth occurs.

In this experiment you will determine the optimum, maximum and minimum pHs for the growth of a bacterium—*Escherichia coli*, a yeast—*Saccharomyces cerevisiae*, and a filamentous fungus—*Mucor* sp.

PROCEDURE

You will be provided with tubes containing 10 cm³ of sterile buffer solutions at pH 2·0, 3·0, 4·0, 5·0, 6·0, 7·0, 8·0, 9·0, and 10·0.

    1. Take an 18 h nutrient broth culture of *E. coli* and an inoculating loop.

2. Obtain a tube of buffer for each of the pH values, and also nine tubes, each containing 10 cm³ of double strength nutrient broth.
3. Observing aseptic technique mix the pH 2·0 buffer solution with the first tube of nutrient broth. Mix by rolling between your palms. Label pH 2·0.
4. Repeat Stage 3 for the remaining pH values and label each accordingly.
5. Flame-sterilise an inoculating loop and cool. Take up a loopful of *E. coli* suspension and using aseptic technique, transfer to the pH 2·0 tube. Label *E. coli*.
6. Repeat Stage 5 and inoculate the remaining eight tubes. Label with your Group No. and Experiment 7.2.
7. Incubate your tubes overnight at 37°C.
8. Take a slope culture of *S. cerevisiae*. Take also nine more tubes of sterile buffer solution and nine tubes each containing 10 cm³ of sterile 4% malt extract.
9. Using the procedure carried out in Stages 3 and 4 mix the buffer and malt extract, and label the tubes accordingly.
10. Flame-sterilise your inoculating loop and cool. Pick up a small amount of the *S. cerevisiae* culture and inoculate the first tube. Label *S. cer.*
11. Repeat Stage 10 and inoculate the other eight tubes with yeast.
12. Incubate for 48 h at 25°C (or at room temperature).
13. Take a slope culture of *Mucor* sp., a Pasteur pipette and some sterile distilled water. Collect nine tubes of sterile buffer solution and nine sterile Petri dishes.
14. Add a small amount of sterile distilled water to the slope culture of *Mucor* sp. and shake to produce a spore suspension.
15. Using the sterile Pasteur pipette take up some of the spore suspension. Add one drop to each of the nine Petri dishes. Label each dish Experiment 7.2, Group No. . . . , *M*. sp. and one of the pH values.
16. Collect a tube of double strength malt extract agar (MA) from the 45°C water-bath. Add the 10 cm³ of pH 2·0 buffer solution to the MA. Mix and pour into the first Petri dish, i.e. the one that you labelled pH 2·0. Rotate to mix the medium and inoculum.
17. Repeat Stage 16 to mix, and pour the other eight plates.
18. Incubate at 25°C (or room temperature) for 3 days.

ASSESSMENT

19. After incubation collect your tubes/plates. Using an arbitrary scale, e.g.

$$0 \xrightarrow{\hspace{3cm}} ++++$$

no growth   very dense growth

Describe the growth which has occurred in the tubes/plates.
20. Tabulate your results and record the optimum, maximum and minimum pHs for each organism.

QUESTIONS

1. Can you make a general comment on the optimal pHs of the various organisms.
2. How would your results influence your choice of a laboratory medium suitable for culturing these organisms?
3. Can you correlate the effect of pH on growth with any methods of food preservation?
4. How do you think pH can influence the growth of the organism?

EXPERIMENT 7.3

THE EFFECT OF OSMOTIC PRESSURE ON THE GROWTH OF MICRO-ORGANISMS

THEORY

One environmental factor which may influence the growth of micro-organisms is osmotic pressure. Practically no cell takes up all solutes, and so the membranes are described as being **selectively permeable**. In theoretical discussions, an ideal membrane, which is impermeable to all solutes, but permeable to water (solvent) is used, and is called a **semi-permeable membrane**. No biological membrane is strictly semi-permeable but cell membranes behave as semi-permeable membranes over short periods of time. This is due to their relatively greater permeability to solvent (water) than to solute.

Osmosis is the net diffusion of solvent molecules through a semi-permeable membrane. If you separate water containing dissolved solutes, from pure water, by means of a semi-permeable membrane, then water passes from the pure water side through the semi-permeable membrane into the solution. This will continue until the concentration of water is the same on both sides of the membrane. If the membrane were permeable to both sets of molecules, i.e. water and solute, equilibrium would be reached by both solvent and solute molecules diffusing through the membrane. *Figures 7.2a* and *b* show these situations.

In the following discussion we will use the term 'semi-permeable' and simplify the situation by considering water, to which membranes are permeable, and only solutes to which membranes are impermeable.

If the water and solutions are mounted in an apparatus as in *Figure 7.3*, and pressure applied via the piston, the flow of water can be prevented. The pressure you apply is the **osmotic pressure** of the solution, i.e. it equals force *P* in *Figure 7.3*.

● = Solvent, e.g. water molecule   ○ = Solute molecule

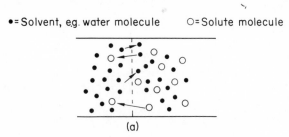

(a)

Figure 7.2 (a) *Membrane permeable to both molecules*

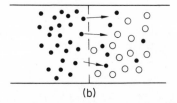

(b)

Figure 7.2 (b) *Membrane semi-permeable*

The tendency for water to pass through the membrane depends on the total concentration of the dissolved particles in the solution. At standard temperature and pressure a molar solution (either molecular or ionic) has an osmotic pressure of 22·4 atm, i.e. you must apply a force of 22·4 atm pressure (2350 lb/in^2) to the situation in *Figure 7.3*, to prevent diffusion of water occurring. Thus a molar situation of sucrose has an osmotic pressure of 22·4 atm, but a molar solution of sodium chloride has an O.P. of 44·8 atm. This is because each ion, for example Na$^+$ and Cl$^-$, exerts its own osmotic influence. (In practice this is not strictly true since only ideal solutions completely dissociate.)

As already noted, the living cell membrane over short periods of time, behaves as a semi-permeable membrane, and therefore will be affected by the O.P. of the environment. If cells contain dissolved materials in greater concentration than the environmental situations, then water will diffuse into the cell. In this situation the environment is **hypotonic** to the cell. If, however, the solute concentration of the environment is higher than the cell, i.e. **hypertonic**, then water

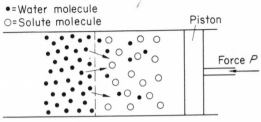

• = Water molecule
O = Solute molecule

Piston

Force P

Force P = O.P. of solution

Figure 7.3 *To show an Osmometer*

diffuses out of the cell and it becomes dehydrated. If the cell and the external solution have the same osmotic pressure, they are **isotonic**.

In this experiment you will investigate the osmotic effect the environment has on the growth of a number of micro-organisms.

PROCEDURE

1. You will be provided with broth cultures of the following organisms:

   Yeasts: *Pichia membranefaciens*
             Saccharomyces cerevisiae
             Saccharomyces rouxii
   Bacteria: *E. coli*
             Halobacterium salinarum

2. Take five sterile tubes of 2% malt extract which contain the following range of glucose solutions—0%, 10%, 20%, 40% and 60%.

3. Take the broth culture of *P. membranefaciens*. Flame-sterilise an inoculating loop and allow to cool. Inoculate the first tube using aseptic technique. Label the tube *P.m.* and 0% g. Repeat for the remaining four tubes and label them *P.m.* and with the glucose concentration.

4. Repeat Stages 2 and 3 using *S. cerevisiae* and then *S. rouxii*.

5. Now take six sterile tubes containing 2% malt extract, and the following range of sodium chloride solutions—0%, 2·5%, 5%, 10%, 20%, and 30%.

6. Inoculate them (as in Stage 3) with *P. membranefaciens*. Repeat with the other two yeasts. Label with the organism and concentration of NaCl.

7. Now take five tubes containing sterile nutrient broth and glucose solutions—0%, 10%, 20%, 40% and 60%, (i.e. one tube for each concentration). Inoculate these with *E. coli* and label accordingly.

8. Repeat Stage 7 using *H. salinarum*.
9. Take six tubes of sterile nutrient broth plus the range of salt solutions 0–30%. Inoculate with *E. coli* and label each tube. Repeat with *H. salinarum*.
10. Incubate all your yeast tubes at 25°C for 48 h. Incubate *E. coli* tubes at 37°C, for 24 h and *H. salinarum* at 37°C for 7–14 days.
11. Assess the amount of growth produced on a nominal scale, for example $0 \rightarrow + + + + +$. Calculate the approximate osmotic pressure for each of the solutions used, (ignore the malt extract and nutrient broth) and construct a table of results as in *Table 7.2*.

Table 7.2 RESULTS TABLE FOR EXPERIMENT 7.3

Organism	Growth										
	\multicolumn{6}{c}{Salt, %}				\multicolumn{5}{c}{Glucose, %}				
	0	2·5	5	10	20	30	0	10	20	40	60
			OP *in* atm						OP *in* atm		
P. membranefaciens											
S. cerevisiae											
S. rouxii											
E. coli											
H. salinarum											

QUESTIONS

1. What do your results indicate with regard to the effect of osmotic pressure on the organisms used? Can you explain these effects?
2. Is there any difference between the effect of O.P. due to sugar, and the effect of O.P. due to salt on any of the organisms used? If so, explain why you think this is.
3. Why do you think bacteria and yeast cells do not normally undergo lysis (osmotic bursting) in dilute solutions? (c.f. red blood corpuscles.)
4. *Pichia, Saccharomyces cerevisiae*, and *E. coli* are more typical organisms than the others used, with regard to the influence of O.P. on their growth. How is this influence used in the preservation of food materials?

EXPERIMENT 7.4

AN EXAMINATION OF WATER SUPPLIES FOR COLIFORM ORGANISMS
THEORY

Water is a medium by which diseases such as typhoid fever, dysentery and cholera can be spread from one human being to another, mainly via drinking water. Water used for human consumption may become contaminated with human faecal material from sewage, and therefore strict bacteriological control of water supplies is necessary.

Normally, **pathogenic** (disease-causing) organisms, if present, are in such low numbers that a comprehensive examination of water for these organisms would be both time-consuming and too slow to yield results of practical value. Tests are carried out, however, for the presence of certain bacteria that occur in larger numbers in human faeces. If these bacteria are present, then it is virtually certain that the water has been contaminated with human faeces, and thus this water presents a possible hazard to the health of the community.

The bacteria tested for are *Escherichia coli* Type 1, *Streptococcus faecalis*, and *Clostridium welchii*, all of which are found in the intestine of normal healthy human beings. Bacteria of the *E. coli* type are called **coliform** bacteria. They are short, Gram negative, non-spore forming, rods, that ferment lactose to produce gas and acids. Coliforms include the Genus *Aerobacter* as well as *Escherichia*.

E. coli (particularly *E. coli* Type 1) occurs in the human intestine, whereas *Aerobacter* occurs typically on hay grains, and other vegetable matter.

In this experiment you will carry out The Presumptive Coliform Test on a water sample, and attempt to isolate *E. coli*. Positive results obtained in the test may not always be due to the presence of coliforms —hence the term 'presumptive'. This test is normally carried out on water which has been purified and on this basis the following classification is used.

Presumptive coliforms/100 cm³	Classification as to potability
Less than 1	Highly satisfactory
Between 1–2	Satisfactory
3–10	Suspicious
Greater than 10	Unsatisfactory

PROCEDURE

You will be provided with a sample of water for testing. Prepare the sample as follows:

(a) Rapidly invert the bottle 25 times.
(b) Pour away one-quarter of the contents.
(c) Shake the bottle vertically, holding the lid tightly, so that the bottle goes up and down through a distance of one foot.

1. Take a sterile 5 oz flat medicine bottle containing 50 cm³ sterile double strength MacConkey's purple broth, and an inverted

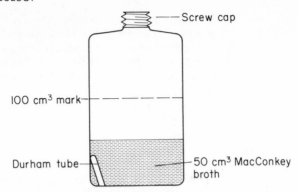

Fig. 7.4 To show a medical flat plus broth and Durham tube

Durham tube. The bottle is marked at 100 cm³ (see *Figure 7.4*).
BE CAREFUL WHEN YOU MOVE THE BOTTLE NOT TO
KNOCK OVER THE DURHAM TUBE.
   (If you do so, then manipulate it to an inverted position again
   and check that it has no air in it.)
2. Very carefully pour 50 cm³ of the water under test into the
   medicinal flat, i.e. pour it in up to the 100 cm³ mark. Again
   beware of the Durham tube. Label the bottle with the Experiment
   No. and your Group No.
3. Take five bacteriological tubes each containing 10 cm³ sterile
   double strength MacConkey's broth and a Durham tube.
4. Using a sterile 10 cm³ pipette, inoculate each of the five tubes
   with 10 cm³ of the water sample. Check after inoculation that
   there are no air bubbles in the Durham tubes. Label the tubes
   with the Experiment No., your Group No., and number them
   1–5.
5. Now take five more bacteriological tubes again sterile each
   containing 5 cm³ single strength MacConkey's broth and a
   Durham tube.
6. Using a sterile 1 cm³ pipette inoculate 1 cm³ amounts of the
   water sample into each of the five tubes. Label them Experiment
   No. 7, Group No., and with the letters a–e. Again check for the
   presence of air bubbles in the Durham tubes.
7. Incubate all your bottles and tubes at 37°C for 24 h.

ASSESSMENT

The broth contains a pH indicator – bromo-cresol purple, and so the
colour of the broth is an indication of the pH of the medium. Tubes
that show a change from purple to yellow have produced acid, i.e.
yellow is acid.

8. Check the Durham tubes for the production of gas. Record tubes that are acid and which have gas in their Durham tubes as positive. Tabulate your results.

9. *Table 7.3* is a statistical table of most probable numbers (MPN table). These tables are used for the enumeration of coliforms in a water suspension. They are based on the assumption that:

   (*a*) the organisms present are evenly distributed throughout the suspension, and

   (*b*) that the organisms can exist in the medium employed in the test, i.e. in the case in MacConkey's broth.

Using the MPN table determine the most probable number of coliforms in the sample. Use the following example to assist you.

EXAMPLE

The following results are obtained from an experiment such as the one you have just carried out

Sample	Results
50 cm³ sample	positive
10 cm³ samples	3 positive/2 negative
1 cm³ samples	1 positive/4 negative

By reference to the MPN table, *Table 7.3*, the most probable number of coliforms per 100 cm³ can be obtained. In this case:

$$MPN = coliforms/100 \text{ cm}^3 \text{ of water}$$

Table 7.3 MPN TABLE FOR COLIFORM BACTERIA

Numbers of tubes giving positive reactions out of:			
1 *of* 50 cm³	5 *of* 10 cm³	5 *of* 1 cm³	*MPN*
0	0	0	0
0	0	1	1
0	0	2	2
0	1	0	1
0	1	1	2
0	1	2	3
0	2	0	2
0	2	1	3
0	2	2	4
0	3	0	3
0	3	1	5
0	4	0	5
1	0	0	1

Table 7.3 continued

	Numbers of tubes giving positive reactions out of:		
1 of 50 cm³	5 of 10 cm³	5 of 1 cm³	MPN
1	0	1	3
1	0	2	4
1	0	3	6
1	1	0	3
1	1	1	5
1	1	2	7
1	1	3	9
1	2	0	5
1	2	1	7
1	2	2	10
1	2	3	12
1	3	0	8
1	3	1	11
1	3	2	14
1	3	3	18
1	3	4	21
1	4	0	13
1	4	1	17
1	4	2	22
1	4	3	28
1	4	4	35
1	4	5	43
1	5	0	24
1	5	1	35
1	5	2	54
1	5	3	90
1	5	4	160
1	5	5	> 180

QUESTIONS

1. What kinds of micro-organisms would you expect to occur naturally in water?
2. Why is water, used for drinking purposes, tested for the presence of faecal organisms?
3. Why are MPN tables used?
4. What is the significance of using three different volumes of media and inocula?
5. What is the purpose of chlorination of drinking water?

EXPERIMENT 7.5

THE MICROBIOLOGY OF MILK

THEORY

Bacteria, both pathogenic and non-pathogenic, may enter milk during its journey down the udder ducts, during milking and handling, and possibly during bottling. The non-pathogenic bacteria come mainly from the soil, air and water, and possibly the udder (picked up from hay, etc.). Pathogenic bacteria come from two sources: from the cow can come the causative organisms of tuberculosis, brucellosis, and mastitis, and from man can come organisms causing diphtheria, typhoid, dysentery and scarlet fever.

To restrict the growth of micro-organisms the milk is cooled immediately after milking and transferred to the dairy in refrigerated containers. Further precautions must be taken, however, and the most common methods are **Pasteurisation** and **sterilisation**. Both techniques involve heat-treating the milk in order to kill pathogens and souring organisms. Treated milk can then be tested in a number of ways, for example methylene blue test, Resazurin test, or colony counts. All three are methods of assessing bacterial activity in the milk.

In the following experiment you will subject raw milk to Pasteurisation and sterilisation and then incubation at different temperatures. You will use the Methylene Blue Test and Colony counts to assess the bacterial activity in your samples.

PROCEDURE

You will be provided with a quantity of raw milk.
1. Take nine 1 oz MacCartney bottles and fill them with raw milk to just below the start of the screw top.
2. Replace the caps and label three of the bottles A, another three B and the remaining three C.
3. Pasteurise the three bottles labelled A by placing them in a water-bath at 63°C so that the water comes to the top of the cap of the bottle. (Therefore ensure that the cap is screwed down firmly.) Leave them for 30 min.
4. Sterilise the three bottles labelled B by placing them in a bath of boiling water and leaving them for 30 min. Again make sure that the water in the bath almost covers the bottles.
5. Leave the three bottles marked C as they are.
6. Take the bottles B out of the water-bath and allow them to cool at room temperature.

7. Take the bottles A out of the water-bath and cool them quickly under a running tap.
8. Divide the bottles into three sets, so that each set has one A, one B and one C in it. Place one set in a refrigerator, one set outside, and one set in the laboratory near a radiator.
9. Put a thermometer with each batch of bottles.
10. Measure the temperature of the three situations daily (or more frequently).
11. Leave the bottles in position for 3 days.

TESTING THE RAW MILK

A. The methylene blue test

(a) Take some of the raw milk and thoroughly mix it.
(b) Take a sterile bacteriological tube and a sterile rubber bung. The tube is marked at 10 cm³.
(c) Remove the bung of the tube and flame the mouth. Pour in 10 cm³ of raw milk leaving one side of the tube unwetted by the milk. Re-flame the tube and replace the bung.
(d) Take a sterile, 1 cm³ blow-out pipette. Take up 1 cm³ of sterile standard methylene blue solution and aseptically add it to your tube.
(*Note:* MAKE SURE the pipette does not touch the milk or the wetted side of the tube.)
(e) Re-flame and replace the sterile bung.
(f) Mix the milk and methylene blue by slowly inverting the tube twice.
(g) Place your tube in a darkened water-bath at 37°C.
(h) At the same time prepare two control tubes using (i) 10 cm³ raw milk and 1 cm³ tap water, (ii) 10 cm³ raw milk and 1 cm³ methylene blue solution. Immerse the two control tubes in boiling water for 3 min. (i) serves to show the complete decolorisation, (ii) to show the colour at the start of the process, i.e. no decolorisation.
(i) Allow your tubes to incubate at 37°C for 30 min. Look at the tube under test from time to time and compare it with the controls.
Note. Make sure you replace the water-bath lid when you have put your tubes in and also after looking at them.
(j) After 30 min remove the tubes and examine them. Total decolorisation of the methylene blue dye is taken to indicate that the milk is unfit for human consumption.
N.B. Decolorisation is regarded as complete when no colour

remains except within 5 mm of the milk surface. Also, a trace of colour at the bottom of the tube can be ignored.

(*k*) Record the result for the raw milk.

B. The colony count

(*a*) Obtain five sterile bacteriological tubes with caps, each containing 9 cm³ sterile distilled water. Label them 1–5.

(*b*) Take the raw milk and using a sterile 1 cm³ pipette take up 1 cm³ of the raw milk and aseptically transfer it to tube 1.

(*c*) Discard the pipette. Rotate the tube between the palms of your hands to mix the contents.

(*d*) Take another sterile 1 cm³ pipette and take up 1 cm³ of the suspension in tube 1, i.e. that prepared above. Aseptically transfer this to tube 2. Again discard the pipette and rotate the tube to mix the contents.

(*c*) Repeat Stage (*d*) to produce three further dilutions using a CLEAN sterile pipette each time.

(*f*) Mix all the dilutions by rotating them again between the palms of your hands.

(*g*) Take five sterile Petri dishes. Label them Experiment 7.5, Group No. . . . , dilution 5, dilution 4, dilution 3, etc.

(*h*) Take tube 5 and the dish labelled dilution 5. Take a sterile 1 cm³ pipette and aseptically transfer 1 cm³ of dilution 5 to the Petri dish. USING THE SAME PIPETTE work backwards through the remaining dilutions, i.e. 4, 3, 2 and 1, and transfer 1 cm³ inocula to the other four Petri dishes.

(*i*) Take five sterile milk agars from the 45°C water-bath. Pour one into each of the five dishes prepared above. Gently rotate each dish to mix the agar and inocula. Allow to cool and set.

(*j*) Invert the plates and incubate at 37°C for 48 h.

(*k*) HANDLE INCUBATED PLATES CAREFULLY AS THEY COULD CONTAIN PATHOGENS. Select the plate which has between 30 and 300 colonies and count the exact number present. (Refer to Experiment 6.1 on page 116 for details of how to count colonies.) Express your answer as numbers of bacteria/cm³ of milk. Collect the results of other groups and calculate a mean result.

TESTING THE TREATED MILKS

12. After 3 days incubation collect your nine bottles. Carry out the methylene blue test on all nine samples, but watch them carefully and record the precise time at which decolorisation occurs.

13. Tabulate your results as shown in *Table 7.4*.

14. Your supervisor will instruct you as to which milk(s) you will deal with. Carry out dilution technique and colony count(s) as shown above. Express your results as numbers of bacteria/cm^3 of milk, and if other groups are working on the samples collect their results and calculate a mean value.

Table 7.4 TO SHOW TABLE OF RESULTS OF METHYLENE BLUE TESTING

Incubation temperature, °C	Milk type	Time for decolorisation, minutes*
(3°C) refrigerator	A B C	
(10°C) outside	A B C	
(23°C) near radiator	. A B C	

*It may be necessary to run the tests for longer than the normal 30 min period.

15. Collect the results for other milk samples and tabulate the results.

QUESTIONS

1. Does Pasteurisation cause a reduction in the bacterial population? If so, how does it do this?
2. How does the efficiency of Pasteurisation compare with the efficiency of sterilisation?
3. From your results say what effect incubation at the various temperatures has on the milk. Explain why you think this is.
4. What does the decolorisation of the methylene blue represent? What causes the colour change?
5. Do the results of the colony counts support the evidence obtained in the methylene blue tests?
6. What do your results tell you about the rate of growth of bacteria in milk?

EXPERIMENT 7.6

TO INVESTIGATE THE ABILITY OF FUNGI TO COLONISE A RANGE OF SUBSTRATES

THEORY

We have already seen in Experiments 1.1 and 6.1, that micro-organisms occur in a variety of habitats. Clearly their ability to exist in a habitat is directly related to their ability to utilise the nutrients available. Are all micro-organisms able to colonise the same substrates with equal success, or are they specific?

The nutrient requirements of micro-organisms (see Experiments 2.1 and 2.2) bear marked similarities to our own, and it is therefore not surprising that some of our foods are also foods for micro-organisms. This is not to our advantage, and so often preservatives and inhibitors of micro-organisms are added to our foods.

In this experiment you will investigate the ability of several fungi to grow on a number of food substrates, and also investigate the random colonisation of these same substrates, by exposing them to the atmosphere.

PROCEDURE

Part 1

1. Take five sterile Petri dishes. Obtain some samples of the following foods: cheese, bread, carrot, potato and orange peel.
2. Using a flame-sterilised scalpel, prepare slices of these foods about 0·5–0·75 cm thick, and cut so that they will fit into the middle of one of the Petri dishes, leaving a 1 cm gap all round.
3. Fold or crumple several filter papers so that they fit into the bottom of the Petri dish and will allow the food slice to rest on top of them (see *Figure 7.5*).

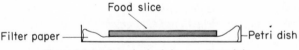

Figure 7.5 *To show the food slice in a Petri dish*

4. Place the slice of bread on the filter paper in the first Petri dish. Using sterile distilled water, soak the filter paper. DO NOT SATURATE THE PAPER OR LEAVE FREE WATER IN THE DISH.
5. Repeat Stages 3 and 4 for four more dishes, and place the other four food slices in position.

6. Expose the dishes in the laboratory for 24 h. Leave the Petri dish lids near to the dishes as shown in *Figure 7.6*.

(*Note.* Part 1 continues at Stage 17.)

Figure 7.6 *To show the position of the Petri dish and lid during exposure*

Part 2

7. Take 20 GLASS Petri dishes and some more samples of the same foods used in Part 1.
8. Cut five slices of EACH food, using the same dimensions as before (Stage 2), and place them into the glass Petri dishes. You DO NOT need filter paper.
9. Place all 20 dishes into an autoclave and autoclave at 10 lb/in² for 10 min. Allow the autoclave to cool and remove the dishes. Allow the dishes to cool.
10. Divide the dishes into four sets of five dishes, containing one of each food type per set. Take one set and a culture of *Mucor plumbeus*.
11. Flame-sterilise a scalpel and use it to cut out a small triangle of the *Mucor plumbeus*, using aseptic technique.
12. Transfer this triangle to the first food dish and lay the inoculum face down so that the hyphae are in contact with the food.
13. Repeat Stages 11 and 12 and inoculate the remaining four foods with *M. plumbeus*.
14. Label the dishes Experiment No. 7.6, and with your Group No., the abbreviation *M. plum.*, and the food type used.
15. Now repeat Stages 10–14 but using the following fungi instead of *M. plumbeus*:

> *Absidia spinosa*
> *Penicillium chrysogenum*
> *Rhizopus nigricans*

16. Leave all the plates to incubate at room temperature for 1 week.

Part 1 (continued)

17. After 24 h replace the Petri dish lids. If necessary, resoak the filter paper pads using sterile distilled water.
18. Return after 3 days and redampen the filter pads if necessary.

ASSESSMENT

Part 1

19. Examine your plates visually and note the extent of growth on them.
20. Attempt to identify as many fungi as possible, using a microscope and reference sources. You can employ two methods of microscopic examination.
    (a) Stereobinocular examination. Using a binocular microscope examine the surface of the fungal mycelium, and a section (slice) of food plus mycelium, laid on its side. In this way gain some idea of the three-dimensional structure of the fungus.
    (b) Wet mounts. Using an ordinary monocular microscope, take a small area of mycelium from the colony on the food. Avoid taking more than a small amount of the food. Mount this on a clean glass slide in water. Examine this mount under low-power and high-power objectives for structural details. Use of stains may be helpful, for example cotton blue in lactophenol.
121. Make a table of your results showing organisms present and the extent of growth.

ASSESSMENT

Part 2

22. Examine the 20 plates. Using a points scale such as the one below tabulate the extent of growth on the various substrates (see *Table 7.4*).

Table 7.4

Fungus	Bread	Potato	Carrot	Cheese	Orange peel
M. plumbeus					
A. spinosa					
P. chrysogenum					
R. nigricans					

Key: + + + + + —overwhelming growth, dish full of mycelium
+ + + + —substrate covered by mycelium
+ + + —growth in patches, quite dense
+ + —growth thin, uneven
+ —little growth

QUESTIONS

1. Which of the foods is colonised more than any of the others? Give reasons why you think this is.
2. Are any of the foods not utilised at all?

3. What factors do you think will influence the colonisation of a substrate by a particular fungus? Relate this to your experimental results.
4. What is the use of the filter paper bed in this experiment?
5. Do any of the fungi used for inoculation of the foods occur in the air flora?
6. On the basis of your experimental results can you answer the question (in the Theory Section), 'Are all micro-organisms able to colonise the same substrates with equal success, or are they specific'?

8
Interaction

INTRODUCTION

In Chapter 7 we considered some of the effects the environment has on micro-organisms. This, however, does not give a complete picture of the situation an organism has to deal with, as other organisms living in the same habitat may have a profound effect on its growth and survival.

The term we use to describe the dynamic relationship between living organisms is **interaction.**

In this chapter we investigate several interacting systems. The first is concerned with the competitive inhibition of one organism by another, via the production of antibiotics (a facility of great value to man). Experiments 8.2 and 8.3 investigate the closer physical relationships of parasitism, in the form of a facultative plant parasite and a bacteriophage virus. Finally, we look at an experiment in which an interacting system leads to mutual benefit, i.e. a symbiotic association.

It may be interesting after completing these experiments to speculate on the evolution of these interacting systems.

EXPERIMENT 8.1

AN INVESTIGATION INTO ANTIBIOSIS

THEORY

Antibiosis is the inhibitory effect one micro-organism has on another micro-organism. The value of this phenomenon commercially is that micro-organisms are used to produce bactericidal chemicals, i.e. drugs, to combat disease in man. Such drugs are called **antibiotics**.

Sir Alexander Fleming in 1928 showed that the fungus *Penicillium notatum* had an antibiotic effect on bacteria. Another species of Penicillium, *Penicillium chrysogenum* is, however, now used to produce penicillin commercially. (Both species of Penicillium are filamentous Ascomycete fungi.) Since penicillin was discovered, many other antibiotics have been found. One of these is streptomycin which is obtained commercially from *Streptomyces griseus*, another filamentous organism, closely related to bacteria.

In this experiment you will investigate the interaction between the three organisms mentioned above, and two different bacteria, *Serratia marcescens* and *Sarcina lutea*.

PROCEDURE

You will be provided with slope cultures (on malt extract agar) of *P. notatum* and *P. chrysogenum*. You will also be provided with a nutrient agar slope culture of *Streptomyces griseus*.

Part 1. Preparing the antibiotic producers

1. Pour two sterile plates of malt extract agar and label MA.
2. Flame-sterilise an inoculating needle (or scalpel), and allow to cool.
3. Using aseptic technique, remove a small triangle of agar plus mycelium, from the slope culture of *P. notatum*. Transfer it to the centre of the first MA plate (put it face down on the agar). Label as Experiment No. 8.1, and with the name of the organism.
4. Repeat Stage 3 with *P. chrysogenum*, and the second MA plate.
5. Pour a sterile plate of nutrient agar, cool and label NA. Flame-sterilise an inoculating needle (scalpel) and cool.
6. Remove a small piece of agar plus mycelium from the slope culture of *S. griseus* and SMEAR it over the centre of the sterile NA plate. Label as Experiment No. 8.1, and with the name of the organism.
7. Leave all three plates to incubate at room temperature for 1 week.

Part 2. Seeding plates with bacteria

You will be provided with a suspension of *S. marcescens* in nutrient broth.

8. Take four MacCartney bottles, each containing 15 cm³ sterile nutrient agar from the 45°C water-bath. Place them in line on the bench.
9. Using a sterile Pasteur pipette take up a quantity of the *Serratia marcescens* broth suspension. Quickly add 5 drops of this to each of the four MacCartney bottles.

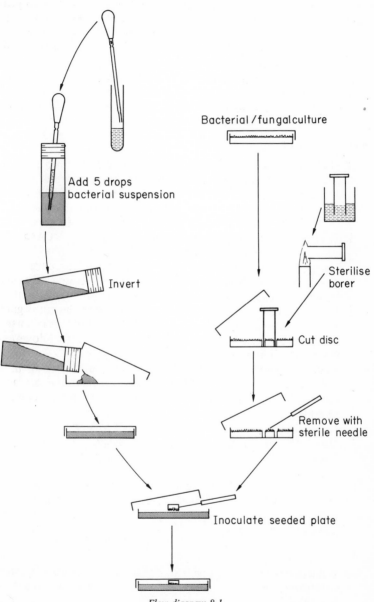

Add 5 drops
bacterial suspension

Invert

Bacterial/fungalculture

Sterilise
borer

Cut disc

Remove with
sterile needle

Inoculate seeded plate

Flow diagram 8.1

N.B. Use aseptic technique when opening the bottles to inoculate.
10. Invert the MacCartney bottles to mix the agar and inoculum, and so produce an even dispersion of bacteria in the medium.
11. Pour the contents of each MacCartney bottle into a sterile Petri dish, thus producing four 'seeded' plates of *S. marcescens* in nutrient agar. Label the plates as Experiment No. 8.1, part 2, *S. marcescens.*
12. You will also be provided with a broth suspension of *Sarcina lutea.* Repeat Stages 8–11 using this *S. lutea* suspension, and so produce four plates of nutrient agar seeded with *S. lutea.* Label them Experiment No. 8.1, part 2, *S. lutea.*

Part 3. Inoculation with the antibiotic producers

13. Dip a cork borer (approximately 1 cm in diameter) into alcohol, and ignite the alcohol by passing the borer through a blue Bunsen flame. Allow the alcohol to burn off, then cool for 30 s.
Note. Only take a SMALL amount of alcohol, and hold the borer so that the alcohol does not run back towards your hand.
KEEP THE ALCOHOL STOCK WELL AWAY FROM THE BUNSEN FLAME
14. Using this sterilised cork borer cut a disc in the culture of *P. notatum* you prepared in Stages 1–3 and remove the cork borer, leaving the cut disc still in the Petri dish.
15. Flame-sterilise a needle and use this to remove the cut disc of *P. notatum* plus agar from the culture dish, and put it on the surface of a 'seeded' plate so that the mycelium is face down on the surface of the agar.
16. Repeat Stages 13–15 using the seeded plate of *Sarcina lutea.*
17. Repeat Stages 13–16 with both *P. chrysogenum* and *Streptomyces griseus.* Add to the label the name of the inoculated organism.
18. Now pour two more sterile plates, one of MA and the other of NA. Re-sterilise the cork borer and using these agar plates cut out circles of agar and place them on the surface of the remaining seeded plates. Label them 'control'.
19. Leave the *Serratia* plates at room temperature for 48 h. Incubate the *Sarcina* plates at 30°C for 48 h (or leave at room temperature).
20. Examine the plates and draw sketches of their appearance. Tabulate your results.

QUESTIONS

1. What do the results suggest with regard to the antagonistic effect of *P. notatum*, *P. chrysogenum* and *S. griseus*, on the two bacteria?

2. How does the result for *P. chrysogenum* compare with the result for *P. notatum*?
3. Why is *P. chrysogenum* used for the commercial production of penicillin, rather than *P. notatum*. (Your answer to Question 2 may give you a clue to this.)
4. Can you suggest ways of increasing the antibiotic production of an organism so that it can be used more effectively for commercial production of the antibiotic?
5. Do you think the production of antibiotics confers any advantage(s) on an organism in its natural habitat.

EXPERIMENT 8.2

SCLEROTINIA—A FACULTATIVE PLANT PARASITE

THEORY

The Ascomycete fungus *Sclerotinia (Monilia) fructigena*, is an organism which parasitises apples causing a disease known as **brown rot**. The disease is of considerable economic importance as it can cause loss of apples both in orchards, during transit, and when stored. Many tons of fruit are destroyed every season as a result of its activities.

The characteristic browning symptom is a *post mortem* change in the apple tissue. It is due to the mixing of a protoplasmic oxidase enzyme with a vacuolar compound released when the cells die. The oxidation of this compound forms a brown-coloured product.

In this experiment you will be provided with a diseased apple, parasitised by *S. fructigena*, and you will use this to study the interaction between a parasite and its host.

PROCEDURE

Part A. Isolating the parasite

1. Using the apple provided, which shows brown rot symptoms, surface-sterilise an area of the epidermis (skin) within the infected area, by wiping it with cotton wool soaked in 1% sodium hypochlorite solution. Leave it for 3 min (see *Flow Diagram 8.2*).
2. Flame-sterilise a scalpel by dipping it in industrial spirit (or 100% alcohol) and igniting the alcohol. Allow to cool and cut a triangular shaped piece from the apple epidermis within the area surface-sterilised. Remove this piece of epidermis.

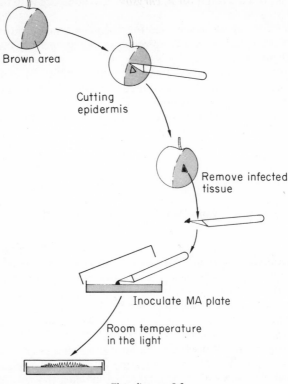

Brown area

Cutting
epidermis

Remove infected
tissue

Inoculate MA plate

Room temperature
in the light

Flow diagram 8.2

3. Re-sterilise the scalpel and remove a small piece of apple tissue from the exposed area (see *Flow Diagram 8.2*). Place this piece of tissue in the centre of a sterile malt extract agar plate. Leave in the light at room temperature for a few days. Describe any growth that occurs.

Part B. Examination of the disease organism

4. Use the apple provided for Part A. Describe as accurately as possible the disease symptoms you can see on the apple.
5. Remove a small piece of tissue from just below the epidermis of the apple (sterile conditions are not necessary). Place the tissue on a slide and add one drop of cotton blue in lactophenol. Gently squash the tissue using a scalpel blade. The cotton blue will stain any hyphae present. Draw some hyphae and try to show their relationship with the host cells.

6. Cut the apple in half. Describe and make sketches to show the extent of the damage, i.e. browning, caused by the fungus.
7. Examine the surface of the infected apple. You should find a number of buff coloured postules on the epidermis. Carefully remove the pustule using forceps and a scalpel, and mount it in water on a clean glass slide. Cover with a cover slip and examine your preparation, and identify and draw a group of asexual spores—**conidia**. These occur in chains where they have erupted through the apple epidermis. Do you think these spores are easily detached, for example by wind or insects?

QUESTIONS

1. When trying to isolate a parasitic organism from diseased host tissue, why should you surface-sterilise the apple?
2. Do the hyphae you have examined penetrate the host cells, i.e. are they **intracellular**, or are they growing only in between the host cells, i.e. **intercellular**?
3. Are the hyphae of this fungus septate or non-septate?
4. How do you think the hyphae manage to penetrate the dense apple tissue? The fact that when the apple is invaded it becomes soft may give you a clue.
5. Fungi which cause plant diseases normally fall into one of two types:
   (*a*) Specialised parasites that are physiologically well adapted to living in contact with the host, but as a consequence have a narrow host range. These organisms usually produce intracellular absorptive structures, called **haustoria**, that absorb nutrients from the living cell. It is thus important that host cells are not killed immediately. As these organisms do not grow saprophytically in the host cells, they are difficult to culture in the laboratory on synthetic media.
   (*b*) Unspecialised parasites that do not produce haustoria. They kill the host cells in advance of the invading hyphae, and then the parasite lives as a saprophyte on the dead tissue. These organisms can be cultured quite easily in the laboratory on synthetic media.
   Which group do you think *Sclerotinia* fits into? Give reasons for your answer.
6. How do you think the disease may spread from an infected apple to a healthy one?

Part C. Theory

Robert Koch in 1876 conclusively demonstrated that the causative organism of the disease anthrax was a specific bacterium. This and other work by Koch, led to a series of tests, called **Koch's Postulates**, which can be used to establish whether or not a particular organism is responsible for a particular disease. There are four major tests:

(a) Is the organism, suspected of causing the disease, constantly associated with the disease symptoms?

(b) Can the organism be grown outside the host in pure culture?

(c) If the organism from the pure culture is inoculated into a healthy individual of the host species, do the typical disease symptoms appear?

(d) Can the suspected organism be re-isolated from the experimental host (as used in (c)), grown again in pure culture, and identified as the original species?

If you were successful with Part A of this experiment, and you have a pure culture of *Sclerotinia fructigena*, then you can carry out test (c) of Koch's Postulates as follows.

1. Surface-sterilise an apple which has no symptoms of brown rot. Use the same technique as in procedure, Stage 1, Part A.
2. Sterilise a scalpel by alcohol dip and flaming, cool, and then cut out a flap of skin from the apple, by cutting three sides of a square and folding the flap back.
3. Cut out a small area of the exposed tissue.
4. Re-sterilise the scalpel and remove a square of agar plus mycelium from the culture of *Sclerotinia* prepared in Part A. *N.B.* Take your square from the edge of the growing colony. Place this inoculum into the hole prepared in the apple and fold back the flap of skin. Place the apple plus inoculum into a sterile beaker and cover with a sterile crystallising dish (or Petri dish lid).
5. Take another uninfected apple and repeat the above procedure but inoculate with a square of malt extract agar only. This is your control.
6. Leave both apples in the light at room temperature for 1 week.

Set up another experiment as follows:

7. Surface-sterilise two more uninfected apples. Cut a small square of agar plus *Sclerotinia* from the edge of your culture, using a sterile scalpel. Place this on the surface of the first apple TAKING CARE NOT TO DAMAGE THE SKIN. Repeat with a piece of malt extract agar only, on the other apple. Again place both apples in separate sterile beakers plus sterile lids. Leave for 1 week in the light.
8. Re-examine all the apples and report on their appearance.

QUESTIONS

1. How do the symptoms produced in these apples compare with those in your original diseased organism?
2. How far do you think Koch's Postulates have been satisfied?
3. Does Part C of this experiment give any clue as to how the fungus gains entrance to its host?
4. In order to control a plant disease efficiently we must have as detailed a knowledge of both its ecology and life history as possible. Apart from the indentification of the organism causing the disease, the following information is of considerable value:
    (a) How does the fungus survive conditions which are not favourable for normal growth, e.g. winter?
    (b) How long can the organism remain dormant and yet still be capable of infecting a host plant?
    (c) What is the host range? Are there commonly occurring plants which can act as alternative hosts? This may be of particular importance.
    (d) Is there more than one host involved in the life cycle? i.e. are two hosts essential for the completion of the life cycle?
    (e) How does the disease spread from one plant to another and when does this happen?
    (f) How does infection occur?

Some of these questions should have been answered by the experiments you have already carried out.

Try to find answers to all the questions. Suggest experiments you could carry out to determine answers to questions not yet solved experimentally. Can you suggest any methods that could be used to control the disease?

EXPERIMENT 8.3

AN INVESTIGATION INTO BACTERIAL VIRUSES— BACTERIOPHAGES

THEORY

Viruses are parasites of living cells that are so specialised that the only activity they show is inside the host cells. Normally their activities damage the host cells, and so viruses are considered to be disease-causing organisms. The first viruses to be described were disease-causing types in plants and animals.

Unlike other organisms, viruses are largely identified by the effects they produce (disease symptoms) in specific animals and plants, and therefore they are not given specific and generic names, but are given names related to the host and the effects of the attack.

In this experiment you will look at the effects of a bacteriophage virus—i.e. a virus that brings about lysis (dissolution) of young, actively growing bacterial cells. The particular virus you will study attacks *E. coli* cells and is known as the T_2 virus.

Bacteriophages are very important because they have served as the principal experimental organisms used for the study of viruses, and have also played an important role in genetical research. This is because the host organisms are relatively easy to handle in the laboratory and the viruses can be grown under controlled conditions.

PROCEDURE

1. Take overnight broth cultures of *E. coli*, *Serratia marcescens*, and *Aerobacter aerogenes*.
2. Take a sterile Pasteur pipette and using aseptic technique remove a small amount of the *E. coli* suspension. Put 5 drops into 10 cm³ sterile distilled water. This gives an approximate dilution of 1/1000.
3. Obtain a previously dried sterile nutrient agar plate. Place 5 drops of the diluted culture of *E. coli* on the surface of the plate using ANOTHER CLEAN STERILE Pasteur pipette. Dip a spreader (L-shaped piece of glass rod) into 100% alcohol and flame—let the alcohol burn off and cool for 30 s. Use the now sterile spreader to spread the *E. coli* inoculum over the plate.
4. Repeat Stages 2 and 3 for *Serratia marcescens* and *Aerobacter aerogenes*.
5. Label all three plates to show the Experiment No., Group No., and the organism used.
6. Allow the plates to dry in the 37°C incubator, but do not invert them; dry as shown in *Figure 8.1*.

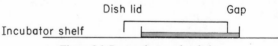

Figure 8.1 *Drying the inoculated plate*

7. You will be provided with an overnight culture of *E. coli* and the T_2 bacteriophage, i.e. a mixed culture.
8. Attach the special sterile filter apparatus to a vacuum pump or tap pump (see *Figure 8.2*). Filter the culture. On completion remove the rubber tube from the pump (note cotton wool to prevent contamination).

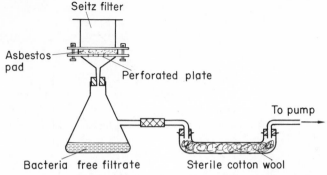

Figure 8.2 *Apparatus for filtration of mixed bacterial/bacteriophage culture*

9. Remove the filter from the flask neck and flame the flask. Pour the contents of the flask into a sterile MacCartney bottle, flame the bottle and replace the cap. Re-flame the filter and dispose into disinfectant.
10. Remove the plates from the incubator. Take a sterile Pasteur pipette and take up some sterile distilled water. Place two drops of sterile distilled water on the plate in the positions shown in *Figure 8.3*. Using the same Pasteur pipette (having emptied it of sterile distilled water, flamed and cooled) place two drops of the filtrate on the plate, again refer to *Figure 8.3* for positions. Repeat this procedure for the other two plates. Incubate the plates at 37°C overnight.
11. Remove a drop of filtrate and make a simple stained preparation. Examine the film with the oil-immersion lens. Make a film of *E. coli* and stain this. Compare the two films.
12. After incubation, examine the plates, looking for clear areas in the rather even growth of organisms on the surface of the plates. Clear areas are known as **plaques** and are evidence of the activity of the bacteriophage virus.

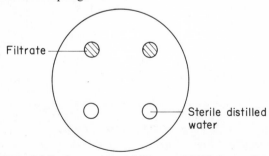

Figure 8.3 *To show positions of filtrate and distilled water drops*

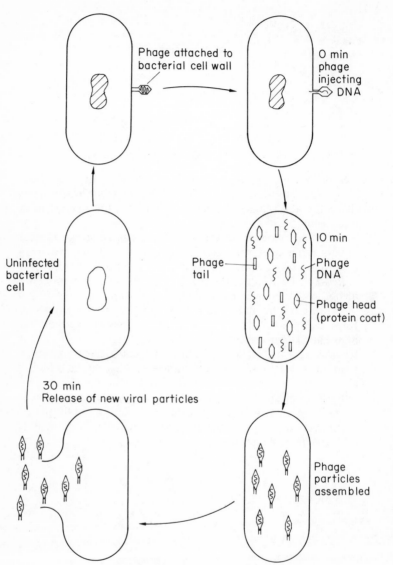

Figure 8.4 *Infective cycle of T₂ phage*

QUESTIONS

1. How are the plaques formed? (See life cycle diagram—*Figure 8.4*—which may be helpful.)
2. Can you reach any conclusion from your experiment about the size of viruses?
3. What do your results show with regard to the specificity of this virus?
4. Can you suggest a method for determining the numbers of virus particles in a suspension?
5. Bacteriophage viruses are vaguely tadpole-shaped with a head made of a protein coat enclosing DNA, and a protein tail. During infection of the bacterium, viral DNA is injected into the bacterial cell (see *Figure 8.5*), and the protein coat is left behind. Mechanical removal of the coat from the cell surface, after DNA has been injected, does not in any way affect future development of the virus. What role do you think the protein coat plays in the life of the virus?

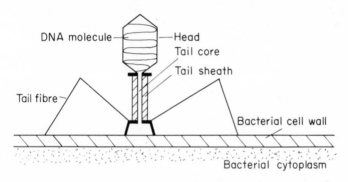

Figure 8.5 (a) T_2 *phage attached to bacterial cell wall*

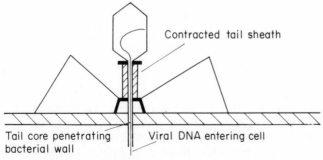

Figure 8.5 (b) T_2 *phage injecting DNA into bacterial cell*

6. Only viral DNA is injected into the cell and yet after about 30 min, many new virus particles are released by cell rupture. How do you think new virus particles are produced?
7. If we assume that a bacteriophage is a typical virus, how do viruses differ from 'normal' cells?

EXPERIMENT 8.4

NITROGEN FIXATION BY A SYMBIOTIC ASSOCIATION

THEORY

If one considers organisms living under natural conditions, a number of situations exist in which organisms of different types live in association with one another. One such association is the symbiotic association, in which organisms live in very close physiological contact to their mutual benefit. The leguminous plant clover, and the bacterium *Rhizobium trifolii* are organisms which exhibit **symbiosis**. The benefit to the host plant (clover) is that when in symbiotic association the bacterium can fix atmospheric nitrogen, i.e. incorporate nitrogen gas from the air into chemicals utilisable by the clover for the synthesis of organic nitrogen compounds. The bacteria obtain a suitable environment for growth and also obtain organic carbon compounds from the host.

It is significant to note that on their own, neither the *Rhizobium*, nor the clover can fix atmospheric nitrogen. This ability to fix nitrogen is a rare one, and thus this association is of real value to both organisms. The association is particularly important under conditions of low available nitrogen in the soil in which the host is growing, as in this situation the symbiotic bacteria supply the necessary nitrogen compounds for the growth of the clover.

PROCEDURE

Part A. Isolating *Rhizobium* sp.

1. Take a fresh clover plant and carefully shake the root system in a beaker of tap water. Continue until you have washed off most of the soil.
2. Examine the clean root system. Look for structures appearing as small lumps on the side of the root—the root nodules (see *Flow Diagram 8.3(1)*. Make a sketch to show the size, shape and position of these nodules on the root system.
3. Select a few, firm, pinkish-coloured nodules. Cut them off from the root using a sharp sterile scalpel. Transfer them to a

MacCartney bottle containing 1% sodium hypochlorite solution. Leave for 3 min. (This surface-sterilises the nodule; i.e. kills any bacteria or fungi on the surface.)

4. Using a pair of sterile forceps, carefully (without crushing) transfer the nodules to a MacCartney bottle containing sterile distilled water. Replace the lid of the bottle and thoroughly shake. Decant the liquid off, and put in more sterile distilled water. Shake and decant. Repeat once more to thoroughly wash the nodules.

5. After decanting the third amount of sterile distilled water from the MacCartney bottle, take a glass rod and dip it into 100% alcohol. Light the alcohol taking care TO HOLD THE ROD POINTING DOWNWARDS (thus preventing burning alcohol from running towards your fingers). Cool the rod.

6. Crush the nodules in the MacCartney bottle using the sterilised end of the glass rod. This should produce a milky fluid at the bottom of the MacCartney bottle.

7. Flame-sterilise an inoculating loop and cool. Pick up a loopful of the milky fluid and streak (see *Flow Diagram 8.3 (1)*) across a plate of the special mannitol-yeast-extract congo-red agar.

8. Leave the cultures at room temperature for 1 week. The *Rhizobium* sp. should appear as flat, watery or mucoid colonies on your plates.

Part B. Examining the bacteria and the nodule

9. Take another loop of the milky fluid produced by Stage 6, and make a bacterial film on a grease-free slide. Flood the prepared film with either methylene blue solution or crystal violet. Allow the stain to remain on the slide for 2 min. Wash with tap water and blot dry.

10. Examine your slide using the oil-immersion lens. Try to find bacteria in the film. In the free living state the bacteria are small motile rods but in contact with the host they become transformed into small swollen or in some cases branched forms called '**bacteroids**' (see *Figure 8.6*).

11. If a stained prepared section through a root nodule and parent root is available, examine this and try to answer the following Are the bacteria intracellular (inside the cells)? Does the nodule have a vascular supply? How do you think the nodule was formed (refer to *Figure 8.7*).

Part C. Inoculating clover with *Rhizobium*

12. Take the mannitol-yeast-extract congo-red agar plate inoculated in Stage 7. Select an isolated colony of *Rhizobium* sp. and remove with a sterile loop.

172

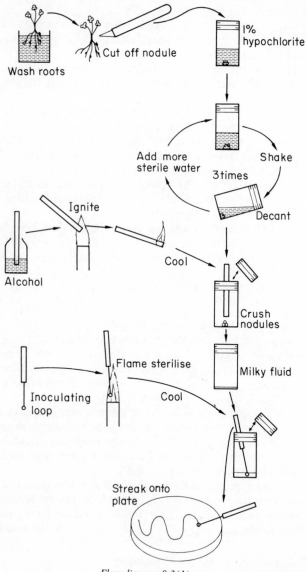

Flow diagram 8.3(1)

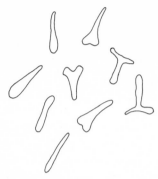

Figure 8.6 *Bacteroids from a root nodule*

13. Place the colony into a MacCartney bottle containing 10 cm³ sterile distilled water. Replace the cap and shake the bottle to produce a bacterial suspension.
14. Put about half an inch of clover seeds into the bottom of a MacCartney bottle, and pour in enough 1% sodium hypochlorite solution to cover the seeds. Leave them for 3 min to surface-sterilise. Wash them with sterile distilled water by shaking and decanting three times as in Stage 4 above.
15. Using a sterile spatula transfer half of the clover seeds to the bacterial suspension prepared above (Stage 13) and shake well. This will inoculate the surface of the clover seeds with *Rhizobium* sp. Keep the rest of the seeds in the MacCartney bottle, having decanted the water.
16. You will be provided with sterilised boiling tubes three-quarters full of vermiculite and plugged with a cotton wool plug. Take six of these tubes and label them 1–6.
17. Take a sterile Pasteur pipette with a wide mouth. Suck up about a dozen seeds from the bacterial suspension and discharge them on the surface of the vermiculite in tube 1. Repeat this for tube 2.
18. Repeat Stage 17 but using uninoculated seeds from the other MacCartney bottle. Place seeds in tubes 3–6.
19. Water the tubes as shown in the table below, using only a few cm³ of solution per tube, and using the SAME amount per tube. DO NOT WATERLOG THE TUBES.

Tube number	Solution used
1 and 2	Nitrogen-free salt solution
3 and 4	Nitrogen-free salt solution
5 and 6	Complete salt solution

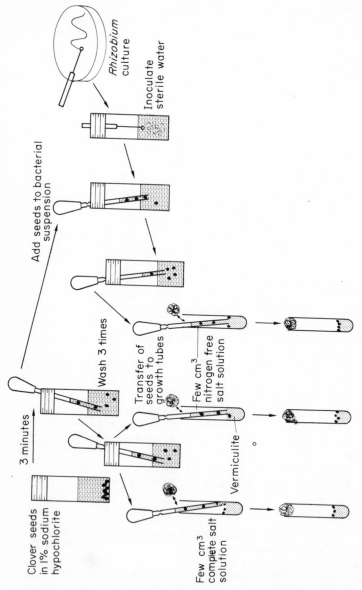

Flow diagram 8.3 (2)

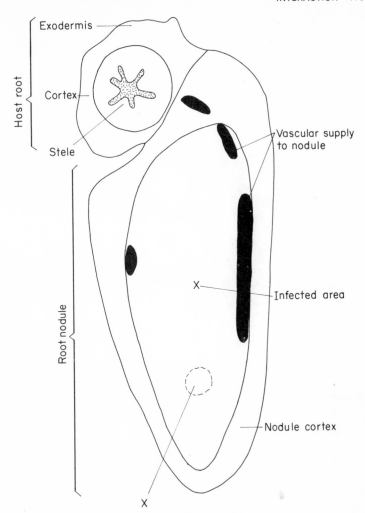

Figure 8.7 (a) *Ts root nodule*

N.B. The salt solution you are using for tubes 1–4 contains all the necessary minerals for growth except nitrate (nitrogen). The complete salt solution used in tubes 5 and 6 has nitrate added.

20. Leave all the tubes in the light.
21. Examine all the tubes at weekly intervals. If necessary, add more nutrient solution. Carefully remove one or two plants and examine them. Record:

(a) The general appearance of the seedlings, i.e. colour, size, etc.

(b) Any change that has occurred since previous examination.

(c) Make a sketch of the plant.

(d) Whether or not nodules have been formed and, if so:

(e) Size, position and colour of nodules.

N.B. The cotton wool plugs can be removed after about 2 weeks.

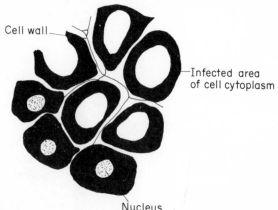

Figure 8.7 (b) *High-power detail of infected area at X on Figure 8.7 (a)*

QUESTIONS

1. What are the possible sources of nitrogen for the clover plants in your experiment?

2. Is there any evidence, from your experiment, that the bacteria are responsible for nodule production?

3. Do your results show that an external nitrogen source is necessary for the formation of healthy clover plants?

4. Is there any suggestion from your results that the association between clover and bacteria is capable of fixing atmospheric nitrogen.

5. Suggest other experimental methods that could be used to confirm nitrogen fixation in the nodules.

6. Explain the important role that symbiotic nitrogen fixation plays in the nitrogen cycle.

7. Why do you think clover can be an important crop?

8. Supposing a leguminous crop which normally produced nodules failed to develop properly in a particular soil. Also, on examination it was found that nodules had not been produced on the plants. What would you conclude from this? How could you solve the problem?

Index